To Jerald Peter

Himself 12·25·76

from Mom & Dad Scafe

THE PILOT'S NIGHT FLYING HANDBOOK

Len Buckwalter

THE PILOT'S NIGHT FLYING HANDBOOK

Doubleday & Company, Inc.

Garden City, New York

1976

Library of Congress Cataloging in Publication Data

Buckwalter, Len.
The pilot's night flying handbook.

1. Night flying. I. Title.
TL711.N5B8 629.132′5212
ISBN 0-385-05460-2
Library of Congress Catalog Card Number 73–9143

Printed in the United States of America

FIRST EDITION

Acknowledgments

The author wishes to thank the dozens of professional pilots, flight instructors, government administrators and people in the aviation industry who helped fill the huge information gap in single-engine night flying. A work on this specialized subject could not have been attempted without benefit of their experience and expertise. The author is grateful to the aviation manufacturers who provided technical and illustrative background material: William Robinson of Beech; Phil Patterson and Jim Shantz of Piper; Jerry Kell and Steve Caine of Cessna; and H. Leverett Jacobi of North American Rockwell. Donald J. Renner of ALNACO provided clear, basic information on airport lighting systems and equipment. Jack Gretta of Whelen Engineering Company skillfully summarized the complex regulations on anti-collision beacons.

The nation's three major FAA centers yielded valuable source material on several medical and legal aspects of night flying. Access to this information at FAA headquarters in Washington, D.C., was made possible through the efforts of Irving Ripps, General Aviation information specialist, Pete Campbell of the Accident Prevention Program, and George Boswell, head of the Airman Certification Branch. Chief of the FAA's Civil Aeromedical Institute in Oklahoma City, Dr. J. R. Dille, supplied the research findings on vertigo. Edward Shoop of the FAA experimental facility (NAFEC) in Atlantic City produced excellent photographs on approach lighting and related systems.

The author is extremely grateful to the considerable assistance given by the National Transportation Safety Board. The NTSB provided him with the first computer printouts which analyze, in great detail, night accidents in single-engine aircraft. They are published for the first time in the Appendix at the end of this book. Gerard M. Bruggink, an NTSB air-safety investigator, also provided fresh insights into the problem of emergency landings in the dark.

The author owes special thanks to Dr. Jesse W. Stonecipher of the Institute of Aviation at the University of Illinois. Through the professor's assistance, the University's unique research program in flare-assisted emergency landings in small aircraft can be described and shown here in great detail.

More than a dozen government agencies and private companies provided valuable illustrations about night flying that appear throughout this book. Their specific contributions are listed below; the remaining photos were taken by the author. ALNACO (24 [bottom], 28, 31, 50 [top]); Beech (17 [bottom]); Cessna (xii, 11, 15, 16, 17 [top], 18, 112, 115); Emergency Beacon Corp. (110); FAA (12, 22, 27 [bottom], 37–39, 43, 44–52, 61, 64 [bottom], 75, 101–4, 117, 119); Flex-Lite Inc. (19); National Weather Service (40 [bottom], 41, 42, 73); North American Rockwell (78, 82, 160); Nova-Tech (108); National Transportation Safety Board (4, 69, 141, 175); Pearce-Simpson (76); Piper (32 [top], 40 [top]); University of Illinois (136, 145–58); Westinghouse (30); Whelen Engineering Co. (13).

Len Buckwalter
Weston, Connecticut

CONTENTS

1	INTRODUCTION	1
2	OUTFITTING THE AIRPLANE	9
	Position lights	11
	Anti-collision lights	12
	Landing light	14
	Interior lights	15
	Flashlights	18
3	LIGHTS AT NIGHT	23
	Rotating beacon	28
	Obstruction lights	29
	Runway lights	30
	Threshold	33
	No lights preceding the threshold	33
	Red lights	33
	Blue lights	34
	VASI	34
	SAVASI	36
	Approach lighting systems	38
	MALSF	41
	MALSR	41
	ALSF-1 and -2	42
	Touchdown zone and centerline lights	44
	Lighting information	46
	Radio control	48

4 VISION AND VERTIGO 53

Taking off into the sunset 54
Flicker vertigo 54
Night vision 55
Illusions 56
Vertigo 60

5 WEATHER AFTER SUNSET 67

Weather at home 72
VHF-FM weather 73
TWEB weather 74

6 FIRST NIGHT FLIGHT 79

Preflight inspection 80
Taxiing 84
Run-up 85
Take-off 86
Landing 89
Simulated emergencies 91
Beyond the pattern 93

7 NIGHTTIME NAVIGATION 95

Compass and pilotage 96
"Night effect" and ADF 97
Compass locators 100
DF steer 101
Surveillance radar 102
Radio failure 105
Repair in the air 106
Emergency homing 107
Which way? 108
Build-and-fade 109
Choice of station 109
Audio volume 109
Ignition noise 109
Portable transmitters 109

8 CROSS-COUNTRY AFTER DARK 113

Preflight tips 116
En route 116
Arrival 118
Instrument proficiency for night cross-country 120

9 ELECTRICAL FAILURE 125

How to read an ammeter 126
Zero-center ammeter 127
Left-zero ammeter 129
Early signs 130
Fuses and breakers 133
How much power remains? 134
Detect the failure as early as possible 134
Turn off every possible electrical item 134
Dead battery in flight 135

10 ENGINE FAILURE 137

Rough engine 138
The descent 139
Touchdown 140
Flare-assisted landing 143
Flare types 146

11 LOG OF A NIGHT FLIGHT 159

APPENDIX: Injuries, accidents which occurred involving single-engine aircraft 173
Cause/factor tables

THE PILOT'S NIGHT FLYING HANDBOOK

Introduction

1

"Slim, whatever you do,
don't get caught in the
air after dark."

The warning was spoken by a professional pilot to a flight student looking forward to his first solo. "A fellow's crazy to fly at night," he continued. "You're up there, you've got to land, and you can't see to do it." But the advice couldn't restrain the young man when months later he had been hired by an outdoor fair to entertain the crowd with aerobatic exhibitions during the day and to release fireworks at night. To meet the terms of his contract, the youth persuaded a dozen drivers to line their cars along the edge of the field so he could land and take off in the beams of their headlights. The time was the Twenties and "Slim" was Charles Lindbergh. It was an era in aviation when night flying was considered an emergency procedure.

Lindbergh's trick with car headlights wasn't new. It had been tried years earlier in the London-to-Manchester race of 1910. Even with its terrors, night flying would determine the winner. The *Daily Mail* had offered a prize of £10,000 for the first man to fly the 185-mile course within a 24-hour period. On April 27, the contenders—Claude Grahame-White of England and Louis Paulhan of France—had covered about half the distance before darkness forced them down. At about 8 P.M. Paulhan retired to a nearby hotel.

With the glory of England at stake, Grahame-White couldn't sleep. He was now 57 miles behind Paulhan and took the daring decision to close the gap by flying at night. At 2:30 A.M. he summoned local motorcycle and automobile drivers to train their acetylene-fired headlights along the field. After take-off, Grahame-White flew nearly two unprecedented hours at night before turbulent winds along the route to Manchester forced him down. The Frenchman wasn't caught napping. Learning of the Englishman's attempt, he took off in the pre-dawn gloom at 4:09 and won the race an hour and a half later. Paulhan said he would not do it again for a hundred thousand pounds.

It was clear that airplanes had to fly reliably at night or remain the impractical contraptions of daredevils and millionaires. Harry Jones tried to prove the point when he made one of the earliest night flights ever recorded over New York City. In 1913 he flew the first parcel post packages from Boston to New York. After midnight, Jones later reported, New York's Great White Way appeared as "a thin streak of brilliant light." The flight ended when he handed ten parcels and a jug of baked beans to the postmaster.

Jones's flight had its flaws. Plagued by weather and engine trouble, the elapsed time between Boston and New York was fifty-two *days*. Soon after his aerial view of Broadway at night, a fierce March wind from the west threatened to blow his biplane out over the Atlantic. He made a forced landing on Flatbush Avenue in Brooklyn and delivered the mail six days later.

Several years later, an Army pilot cast further doubt on whether man would ever fly in the scant visual environment of evening. In a famous experiment he dropped a blindfolded pigeon from an airplane. Maybe men were terrorized by the dark, but birds had navigated the night sky since the Jurassic Period. The pigeon fell helplessly in a tight spiral until it finally held its wings out as an air brake.

The urgency to fly between sunset and sunrise persisted through the 1930s. It grips Saint-Exupéry's classic novel *Night Flight*. In speaking about a night mail service between South America and Europe, the character Rivière says: "It is a matter of life and death for us; for the lead we gain by day on ships and railways is lost each night."

Night flying leaped forward with the U.S. mail. From 1920 to 1927 the Federal Government erected lighted beacons over more than two thirds of the 2,665-mile transcontinental airway system. Blinkers and beacons spaced three miles apart led intrepid pilots through darkness and signaled the mileage to nearby terminals. The safety of night flight zoomed with geometric impact. In seven years, beginning in 1919, pilot fatalities dropped from one in 114,342 miles to one in 2.5 million miles—more than twenty times.

Meanwhile, advances in aircraft kept pace with developments on the ground. A generator driven by a small propeller mounted in the airstream was added to recharge a battery in flight. The growing number of electrical devices led to the modern alternator with a direct mechanical coupling into the engine for steady power, even as an airplane taxis slowly on the ground. Primitive lighted airways were supplanted by hundreds of radionavigation stations for accurate guidance in any direction. By the 1970s, the number of lighted airports in the United States rose to about 4,000 until nearly one out of every three airports in the country boasted at least one lighted runway.

Yet the amount of night flying in light aircraft hardly rivals that of day flight. Aviation magazines often treat the subject with foreboding with article titles like "Fly Single Engine at Night and Live," or "Night Single Engine Yes or No." To find out if statistics confirm the danger of night flying, I queried an accident investigator at the National Transportation Safety Board (NTSB) in Washington, D.C., in 1973. He explained that no one has the answer, because there is no data on "exposure rate." He was referring to night flying's greatest information gap: There is no record of how many hours private pilots spend in the air after dark. Accident rates have always been a matter of speculation.

Nevertheless, anyone in aviation intuitively knows that night flying is riskier than flight during the day. Losing an engine, running into bad weather, finding one's way, or suffering electrical failure are obviously more hazardous after the sun goes down. Although there are no comprehensive night-accident rates, there is little mystery about the leading nighttime killers. They are weather, alcohol and fuel.

Accident reports developed by FAA and NTSB are filled with poignant testimony, like these words of a California well-driller: "I met Mr. ——— about 5:00 P.M. on the afternoon of his plane crash. We had two drinks. I left him about 6:00 P.M. That was the last time I saw Mr. ———."

A mechanic in Pennsylvania wrote: "At approximately 8:45 P.M. EST, I heard the sound of an airplane engine. While the engine was revving up, I heard the sound of a crash, then the engine noise stopped. There was a heavy snowstorm going on at that time."

The other factor in this murderous trio—fuel—is just as shocking. It often refers to running out of fuel in flight because of poor planning.

An illuminating clue about the safety of night flying appears in a table prepared each year by NTSB under the heading "Conditions of Light" (see Fig. 2). Note that night is divided into two categories: dark and moonlight-bright. The surprise is that a pilot is more than eleven times

CONDITIONS OF LIGHT	TOTAL	ACCIDENTS FATAL	NONFATAL
Dawn	36	10	26
Daylight	4031	435	3596
Dusk (twilight)	169	29	140
Night (dark)	466	146	320
Night (moonlight-bright)	42	8	34
Unknown/Not Reported	23	19	4
Total	4767	647	4120

NTSB 1969

2. Most night accidents happen on dark, moonless nights.

more likely to have an accident on a dark night than on a bright night. It is graphic evidence that night flying is, in fact, an intermediate step toward instrument flying. When the moon illuminates the terrain or the horizon stands out in vivid relief, it might as well be day. But during a new moon or a starless, overcast night, a pilot needs help from instruments. An artificial horizon or turn needle must give reassurance when the natural horizon disappears in the pitch-blackness.

Night accidents in single-engine aircraft were broken out for the first time in great detail by an NTSB computer in 1973. A look at this tabulation (see Appendix) reveals that pilots commit much the same errors at night as they do during the day—the stall/spin, continuing flight into adverse weather, and fuel mismanagement. If the engine doesn't stop because of a pilot-induced error or poor maintenance, the chance of power failure is remote.

"Would *you* fly in a single-engine plane at night?" The question was posed to FAA officials, accident investigators, flight instructors, examiners, airline captains and other experienced pilots. Their answers were remarkably unanimous. None would hesitate to fly a well-maintained airplane, or "one that I know," in good weather. One high FAA official simply said that flying at night is not risky "if pilot and plane are qualified."

Aviation insurance companies agree. The underwriters insert numerous limits into their policies but almost never a restriction or rate hike for night flying. In rare cases where a night-flying proscription appears in a policy, it is not a reflection of safety but a marketing ploy. Intense competition between companies turns the night-flying clause into a negotiable item. In nearly all cases, however, the final decision on whether to fly at night remains with the individual pilot or the FBO (Fixed Base Operator) who rents him the aircraft.

There is little debate over night single-engine flight and IFR (Instrument Flight Rules). They don't mix. One insurance company maintains a fleet of single-engine planes to transport underwriters around its territories, and the company sees no danger in night-flying these aircraft. The insurance men have flown several decades with an impressive safety record. There is, however, one strict prohibition: no single-engine IFR night flight. The rule is widely observed throughout general aviation. Pilots willingly fly at night or single-engine IFR, but few do the two simultaneously.

A flying insurance adjuster explains it this way. During a night flight he plans his route over a string of lighted airports. With sufficient altitude, he can make an emergency landing almost anywhere along the way. But on instruments at night he no longer controls the route. Airways are assigned, he is vectored around cities and possibly sent over water beyond gliding distance of land. Now those reassuring airports below are missing. If the engine fails, the airplane may break out of the overcast too low for emergency maneuvers. Low cloud layers would also darken the terrain during IFR conditions and increase the danger.

An FAA flight examiner was concerned about flying solo on instruments in bad weather at night. The cockpit workload can be crushing in the critical moments of an approach. When the aircraft breaks out at the correct altitude and is aligned with the runway the ordeal isn't over. Rain hitting the windshield wipes the approach lights into a shimmering blur.

Clubs and FBOs cope with the weather problem by prescribing ceiling and visibility minimums. A New England operator rents aircraft at night for local flying only when the ceiling is at least two thousand feet and visibility is not less than five miles. These numbers are approximately double for longer, cross-country flight: five thousand feet and ten miles. No flight is permitted if any precipitation is present or forecast. Coupled to these weather minimums are flight areas. An instrument-rated pilot is permitted unlimited flight, while private and commercial pilots can fly up to one hundred and fifty miles from the airport at night. Private pilots with less than a hundred hours as pilot-in-command must remain within seventy miles and use only controlled airports. Students are confined to the traffic pattern. A night checkout is another requirement. Given by an instructor, it is good for sixty days. Beyond that time, a recheck is required if no night flying was done in that period. A large West Coast flying club imposes similar restrictions, but also requires separate checks for local and cross-country flights. Unless a club member has an instrument rating, the check ride must be taken on a moonless night.

Interest in night flying spurted ahead in 1972. An FAA overhaul of pilot certification singled out for the first time the amount of instruction required for night operations. Before then, any private pilot could fly after dark with no prior experience. Only when passengers were carried did the law intercede; the pilot had to log at least five night take-offs and landings in the preceding ninety days. Under the new regulations (effective November 1, 1973), candidates for a private license are given an option of taking three hours of flight instruction at night, including ten take-offs and landings. If the pilot does not log this time, his certificate bears the limitation, "Night flying prohibited." It can be removed at any future time by fulfilling the night requirement.

At the same time, another section of the rules was amended to ease the restriction about passengers. In the ninety days before carrying passengers, the pilot must accomplish only three (it was formerly five) take-offs and landings to a full stop in a period beginning one hour after sunset and ending one hour before sunrise. Further, the old requirement could be fulfilled by flying any airplane. The new rule states that take-offs and landings be done in an aircraft of the same category and class, but not necessarily the same *type*. For example, a pilot may fly a Cessna-150 or a Piper Cherokee-140 to remain current. These aircraft are different types, but in the same category (airplane) and class (single engine).

These sweeping changes in night-flying rules were justified by the FAA under a concept it called "total operational training." It replaced the earlier approach to pilot certification, which depended on passing a written test and demonstrating certain flight maneuvers. Under the new concept a pilot cannot discover night flying on his own, but must receive training under the guidance of a professional flight instructor. (The rules are summarized below.)

Dual Instruction	Three hours at night, including 10 takeoffs and landings	Remarks: Optional requirement to avoid restriction on night flying. Restriction may be removed at any time by taking required instruction.
Night Experience to Carry Passengers	Three takeoffs and landings to full stop at night within preceding 90 days.	Remarks: Aircraft must be same category and class, but does not have to be same type.

3. Summary of night flying requirements for private pilots (effective November 1, 1973).

There are many more reasons to fly after dark besides avoiding the night restriction. Most important, it's defensive flying. Nearly any pilot traveling cross-country—even short distances beyond the local airport—faces the threat of darkness. The eastern sky in December grows ominously dim soon after 3 P.M. Strong winds of winter can lengthen a trip by enough minutes to determine the difference between a day and night arrival.

An airplane's utility soars with night flying. Ground a plane after dark and its value is eroded by fixed costs of insurance, tie-down and periodic maintenance. Costly instruments and avionics of light aircraft lose value when confined to daytime operations. And these are the devices which turn night flying from an uncertain adventure to a safe, routine activity.

Night flying is a superb introduction to instrument flying. Darkness demands more reliance on instruments, but under far less threatening conditions. The instrument student at night discovers air-traffic facilities less loaded by commercial traffic. Controllers are friendlier and more willing to allow practice instrument approaches.

Night flying is a beautiful experience. The air is often smooth and clear. Few aircraft crowd the sky. Cities and towns glitter in unending fascination to deepen the mystery of flight that touches every pilot. With a reliable craft and careful piloting there is little fear of getting "caught in the dark."

Outfitting the airplane

2

Certain struts . . . and other parts of aircraft are to be made sufficiently luminous at night by being covered with luminous paint to assist in night flying.

ROYAL AIR SERVICE MEMORANDA, 1915

Today's light aircraft often carry an electrical system of more than 500 watts, a dozen lamps and enough electronic aids to total half the cost of the airplane. Yet FAA regulations require little more than wing and tail lights for night flying. For most flights the rules don't call for instrument lights, a landing light or the most indispensable item of all—a flashlight. Not until 1972 were all civil aircraft required to carry anti-collision lights for night operations. Yet night flying is much safer and more convenient with the optional accessories.

For flight at night, the regulations begin with the instruments of daytime VFR (Visual Flight Rules) shown in Figure 5. The added night equipment includes position lights, anti-collision lights and an adequate

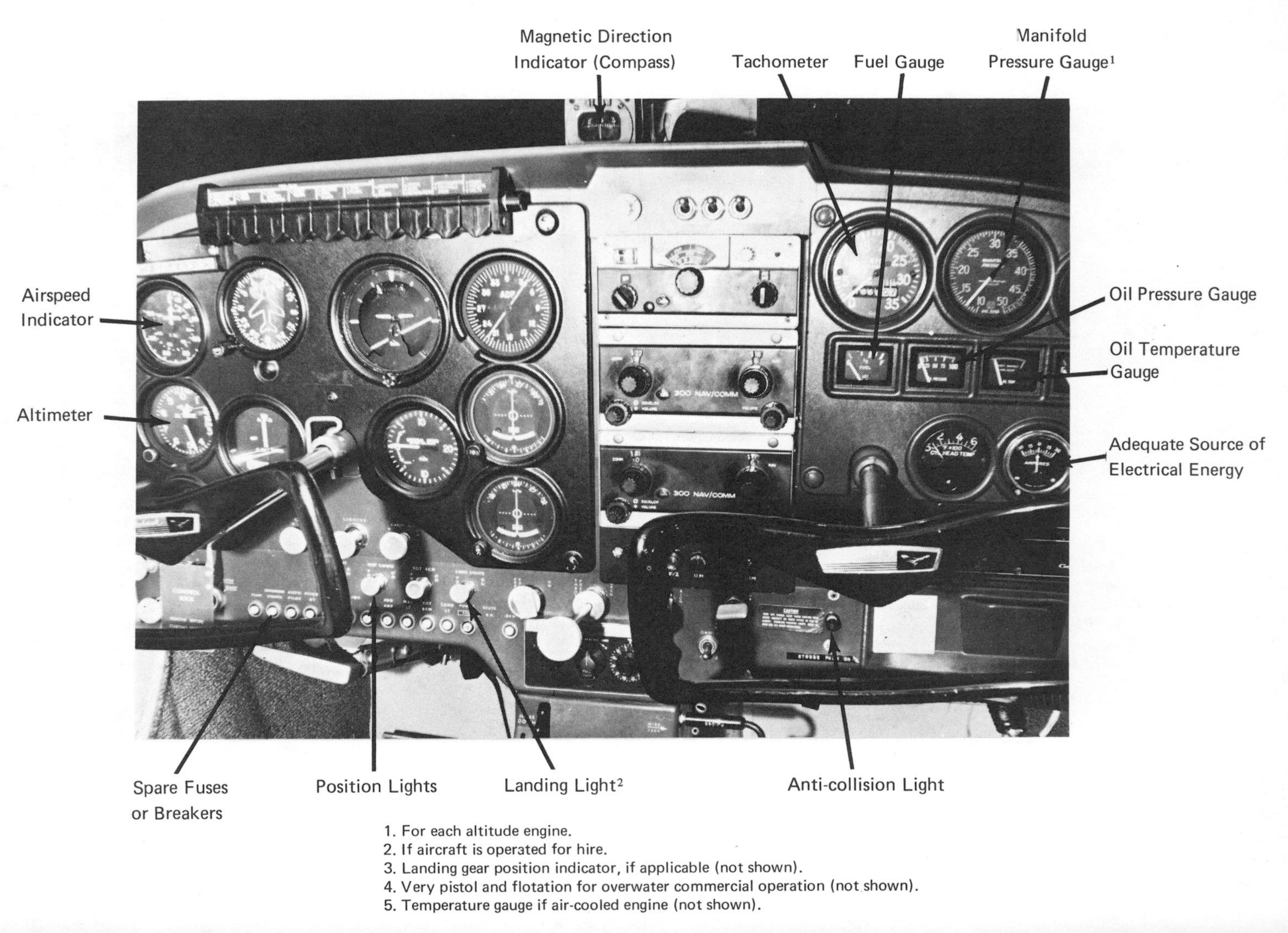

5. Instruments and equipment for night VFR operation. Instruments along top are also required for day VFR.

source of electrical energy for all electrical and radio equipment aboard the aircraft. A spare set of fuses or three spares of each kind are required. (Circuit breakers, which can be reset manually, appear in newer aircraft.) Only airplanes operated for hire must carry a landing light.

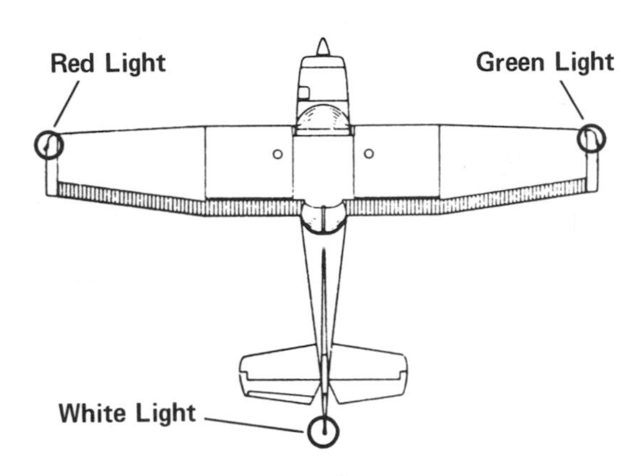

6. Aircraft position lights.

Position lights

Airplanes borrow the marine practice of marking the craft's left side with a red light, the right side with a green light, and the tail with a white light. They're designated "position" lights but are often called "navigation" or "running" lights. It takes little imagination to see that an aircraft, otherwise invisible in the night sky, reveals its relative position by lights. A single white point warns of an aircraft directly ahead. Movement in that direction is difficult to perceive, but a single light arouses several possibilities. If the aircraft ahead is slow, you may be closing the gap; if a fast aircraft has aligned itself with your flight path it could be trailing dangerous turbulence. In either case, the absence of red and green lights means the aircraft ahead is pointing its nose away from you. Masks placed inside position lamps confine the light to achieve the directional effect. Several other possibilities are shown in Figure 7.

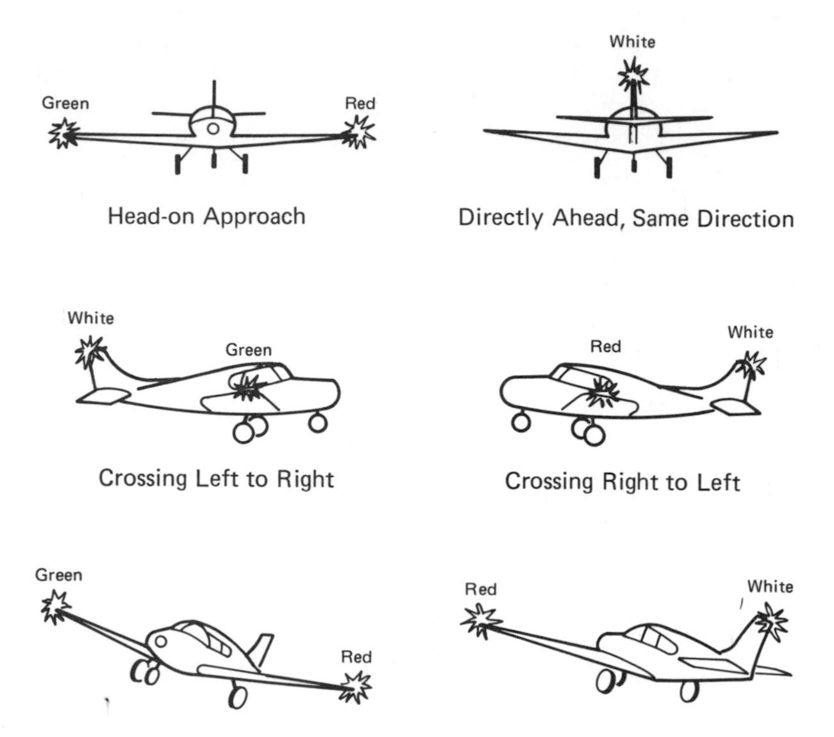

7. Position lights reveal location and heading of other aircraft.

If you see red and green lights only, be ready to take evasive action, because another aircraft is heading directly toward you. Now the white tail light is blocked from view. Another important warning created by position lights is the apparent absence of movement through the sky. It may mean another aircraft is on a collision course with you, especially if it's at your altitude. When motionless position lights appear to your left or right, be extremely wary until they cross your aircraft's nose.

Anti-collision lights

On August 11, 1972, the FAA ruled that in addition to position lights, all powered aircraft must carry anti-collision lights. Most earlier aircraft were equipped with a rotating beacon or flashing incandescent lamp, which continued to be acceptable under the 1972 law. New aircraft which received their type certificates after August 11, 1971, however,

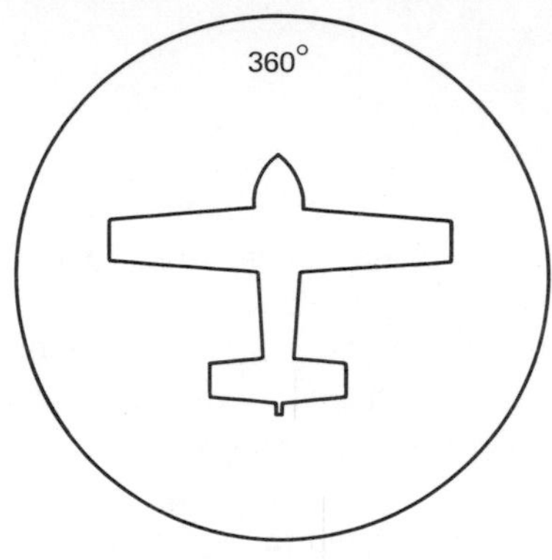

(A) An approved anti-collision strobe light system must project light 360° around the aircraft's vertical axis. One or more strobe lights can be used.

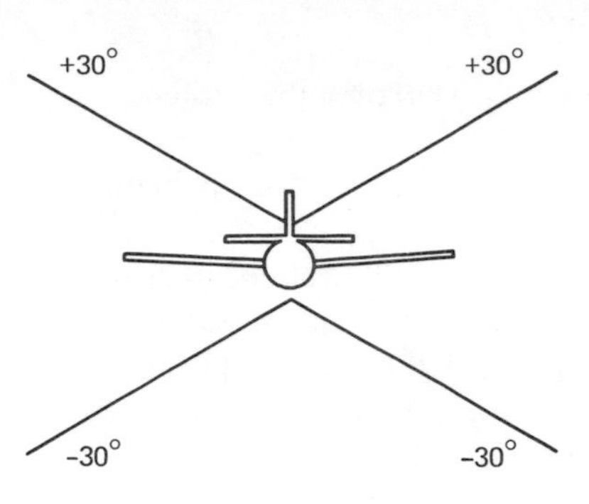

(B) An approved anti-collision strobe light system must project light ±30° above and below the horizontal plane of the aircraft. One or more strobe lights can be used.

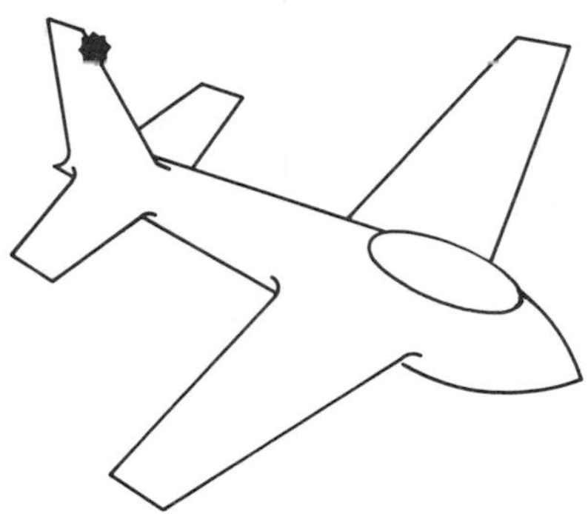

(C) One anti-collision strobe light mounted on the vertical fin will meet the minimum requirements on most aircraft. A half red and half white lens is recommended.

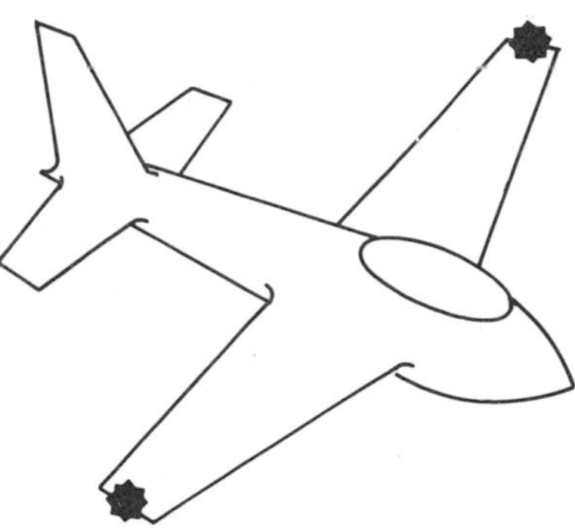

(D) Two wingtip strobe lights that protrude beyond the wingtip, their light converging in front and back of the aircraft within 1200 feet, are considered an approved anti-collision strobe light system.

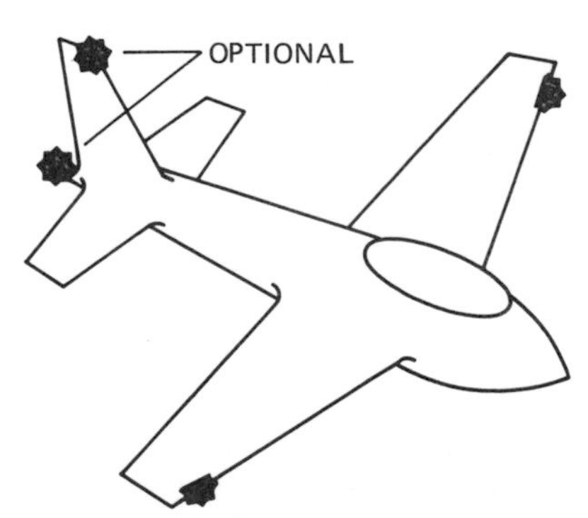

(E) Enclosed wingtip anti-collision strobe lights require a third strobe light on the tail or vertical fin to fill in the required light envelope. This is an approved anti-collision system.

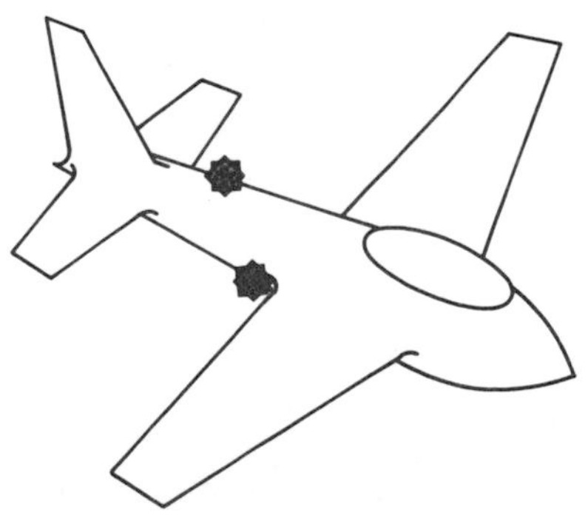

(F) A fuselage-mounted anti-collision strobe light system requires a minimum of two strobe lights to get the required vertical coverage. This is an approved anti-collision system.

8. Anti-collision lighting systems. (Courtesy Whelen Engineering Co.)

must have anti-collision lighting of at least 400 effective candlepower, or about four times greater than the brightness of earlier beacons. The light projects a full circle of light horizontally around the plane in order to warn aircraft closing at the same altitude. In most installations, an anti-collision light atop the rudder fin satisfies a coverage which must extend 30 degrees above and below the horizontal plane. Several other mounting possibilities are shown in Figure 8.

An anti-collision light may be either red or white. Since a red lens reduces the light output, some pilots replace a red lens with a clear lens for greater visibility. High-intensity strobe lights, popular for many years, may also replace an incandescent lamp inside a beacon fixture for brighter light, but at higher cost and complexity. A common arrangement is a strobe mounted atop the rudder fin with a red lens covering the lamp's forward section and a clear lens at the rear. In this "split lens" arrangement there is less disturbance to the pilot's vision, since the full intensity of a white strobe is emitted toward the rear. An intermediate model between the strobe and incandescent lamp is the iodine-quartz element (a type of incandescent). Cost is higher than for a conventional filament lamp because of solid-state flasher circuits.

Strobe lamps radiate electronic interference which might disturb reception on ADF (Automatic Direction Finder) receivers. If noise is heard while tuning ADF bands, and it threatens safe navigation, it should be cured by a competent avionics technician.

An anti-collision light—especially a rotating type—can give a pilot a bout of vertigo followed by possible loss of orientation. It can happen if the beacon remains on while flying in clouds, fog or haze. (What to do about vertigo is described in a later chapter.) Aircraft are placarded with a warning to turn off the anti-collision light under these conditions to prevent a confusing whirl of lights from reaching the pilot's eye.

Landing light

This is an optional accessory in many aircraft and not mandatory for night flight. If the light is installed a pilot should know how to take off and land *without* it, because the lamps are surprisingly short-lived. Typical models are rated for an average life of either ten or twenty-five hours. Nevertheless, the landing light is a clear necessity. Taxiing a half mile to a runway at night over an invisible blacktop is uncomfortable and hazardous. A landing light illuminates obstacles or ditches along the way, and shines a warning to people walking on airport grounds.

9. Landing light mounted under prop spinner.

Some aircraft have a two-position landing-light switch. The first detent turns on one lamp for high-intensity lighting during take-off and landing. The low position relieves the drain on the battery while taxiing and reduces glare to other pilots in the area.

Interior lights

There is little standardization in the cabin and instrument lighting of small aircraft. Factory-installed lights are often minimal and leave critical areas poorly illuminated. This makes a flashlight a practical necessity to fill in the dark corners of most planes. In some aircraft, illumination is blocked by the instrument panel, or light beams fail to strike control knobs or levers mounted near kick pads or below the panel.

The *flood,* simplest form of panel lighting, is a red lamp mounted on the cabin roof and aimed at the instrument panel. Sometimes a second flood is mounted on a post near the pilot's head, with a switch to select red or white. The red beam enhances the overhead flood illumination, while the white light is for map reading. Some aircraft locate a blue-white flood under an anti-glare shield where it casts a general illumination on the panel instruments.

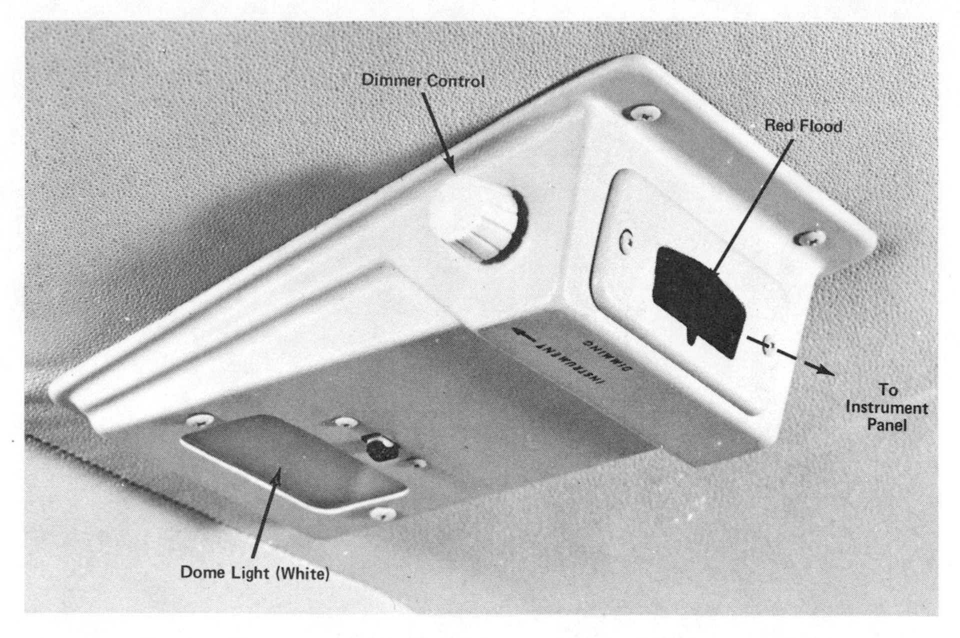

10. Red floodlight on cabin ceiling illuminates instrument panel. Note rheostat knob for dimming.

Eyebrow lights place illumination closer to individual instruments. The lamps are inside a housing which arcs around the top of each instrument. *Post* lights are small, lamp-carrying protrusions at the upper edges of each instrument. *Integral* lights are completely housed within the instrument and provide a uniform glow over the dial face.

Instrument lighting is usually adjustable to desired intensity by a rheostat, or dimmer. Setting lights to a low level helps the eye adapt to darkness or reduces annoying reflections from cabin windows or other shiny surfaces. There is often a second rheostat to independently control illumination to radios and magnetic compass. Radio dials are often internally lighted because their small numbers are too difficult to see under a floodlight, while the magnetic compass is usually mounted beyond the field of the floodlight.

A troublesome light source is the orange reply light of some transponders. The repeating flash annoys the eye, even when the rheostat dims the radio lights to a comfortable level. A bit of tape placed on the lens to cover part of the glow can cure it.

There are several lighting options for the buyer of a new aircraft. Courtesy lights shine beneath a high wing to help passengers alight, small plastic devices glow on wingtips to signal that navigation lights are working, and map lights recessed under a control wheel conveniently illuminate a chart. The most important optional accessory in night flight, though, is purchased at a local store.

11. **Lighting by post lights.**

12. **Instruments with integral lights.**

13. Map light and dimmer control mounted under control wheel.

Flashlights

There is no requirement for a private pilot to carry a flashlight aboard his airplane for night flight. But if a missing gas cap causes an accident, an investigator may ask, "How was a preflight inspection accomplished?" The flashlight is not only necessary during the walk-around, but an item of survival in an electrical failure. You'll reach for a flashlight repeatedly to illuminate the corner of a chart or poorly lighted knob. In a test report on one general aviation plane, for example, *Aviation Consumer* wrote:

> During night operations we found the engine gauges to be inadequately lighted by the airplane's post lighting system. Also, you have to play blind-man's buff with the fuel selector. There certainly should be a fuel selector light since the control is located on the floor underneath the center part of the instrument panel, far enough back that it isn't illuminated by the panel lights. We believe that a flashlight is a no-go item for all night flights or emergencies. In most general-aviation planes . . . a flashlight is necessary for normal operation.

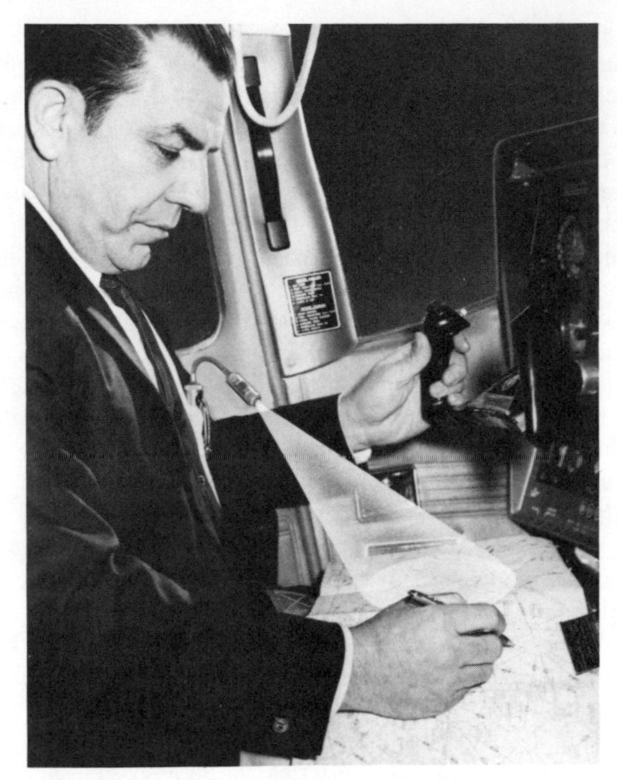

14. Pocket-type flashlight for hands-free operation.

Other pilots would amend that last remark by saying *two* flashlights are necessary for routine operation. Among the more than 150 flashlight models on the market, there is no single type that will cover all the exigencies of night flight. Carrying two flashlights also reduces the distressing possibility of being trapped aloft with no light because of dead batteries, bad bulbs or a corroded switch.

The most familiar flashlight is a two-cell type with D-size batteries and a PR2 bulb. It is fine for preflighting an airplane and lighting the instrument panel when power fails. If a second flashlight is carried, a pocket type with penlite cells is handy, especially a model like the one shown in Figure 14. It clips in the pocket and directs a shaft of light toward charts or instruments in hands-free operation. A red filter on some models prevents any effect on the pilot's night vision. Since red light causes certain features printed in red (or magenta) on an aeronautical chart to disappear, white light is also available.

In selecting a flashlight consider these additional features: *Never* use a flashlight with a magnetized holder in an airplane, because it can wreck the accuracy of the magnetic compass. (Many such flashlights, in fact, are uncomfortable to grasp in the hand.) Before purchasing a flashlight, check to see if there is a small spring between the bulb and battery. In this design there is much less chance for the bulb to break if the light is dropped. Another bit of protection is a lens which is recessed at least one-quarter inch. A desirable feature is a removable cap at the unlighted end of the unit; it's easier to change batteries and there's less stress on the switch contacts at the other end. That switch, incidentally, should have considerable friction to prevent the light from being switched on while bouncing around a flight case. (After several such mishaps, one pilot tapes the switch in the off position.) Bulbs go dead with no warning, so buy a flashlight that stores a spare inside or, at least, carry a spare. If a new bulb is dim, don't assume the cause is a weak battery. Identical flashlight lamps, even when new, vary widely in brightness. The electrical contacts in flashlights will corrode, so clean them occasionally with fine sandpaper.

As any Boy Scout knows, a familiar flashlight design has an L shape. That's an official model, because it can be clipped onto the belt for carrying or for illumination while walking. It is also an excellent type for the night pilot. If aircraft power fails, the flashlight can be held between the thighs so that the light shines directly at the instrument panel and leaves the hands completely free for flying the airplane. (This technique is covered in detail in Chapter 6, "First Night Flight".)

The unpredictable life of batteries is another reason for carrying two flashlights. If a fresh D-size battery is installed in a flashlight, it is typically rated to last eighty days if discharged once a day for five minutes. A dry cell, however, leads a tenuous life. Long storage on a dealer's shelf (sometimes for years) and storage at higher-than-room temperature are common causes for failure. It is not unusual to buy a dead or nearly depleted cell from "new" stock.

Since the flashlight on a night flight is so crucial, consider these buying tactics to avoid poor cells. First, deal with a store with high turnover. If there's no local outlet, your chances of getting fresh cells are better from a large mail-order electronic house than a corner drugstore. If a local store has a battery tester, this can help eliminate poor cells. Your flashlight does this too by showing a yellow, rather than white, light. Don't use mercury batteries; they last longest, but are unsuitable for the heavy currents drawn by a flashlight. Alkaline cells have higher current capacity than standard cells, but are not particularly suited to intermittent flashlight service. Alkalines withstand cold weather well, but also suffer a tendency to leak if not used in several months.

When all factors are considered, a conventional dry cell—the carbon-zinc battery—appears to be the best choice. Within this category are models labeled "extra long life" or with some other intimation of higher capacity. These premium grades are recommended. Avoid any battery or flashlight which makes an extraordinary claim, e.g., "You never have to replace the battery!" There is a perverse truth in the claim: When the battery dies, you throw away the whole flashlight.

Recharging a carbon-zinc battery is not recommended. Although inexpensive charging equipment is widely available and the procedure endorsed by the National Bureau of Standards, it is a tricky process. In exchange for considerable bother, there is only slightly extended battery life. More important is the fact that a recharged dry cell may fail at an unpredictable moment. It is wiser to purchase high-quality carbon-zinc cells, keep flashlight parts clean, and replace the cells if they ooze or the light appears yellow. If you do not expect to use the flashlight for a month or so, remove the batteries to protect against leakage. Then place the batteries in a plastic bag and put them in a refrigerator or freezer. Industry tests prove that dry-cell batteries stored near 0° F. last many years without deteriorating.

Lights at night

3

Powerful visual cues reach out to the daytime pilot. As he flies toward home, familiar contours on the terrain feed his eye and guide him to the airport. Five miles from landing, summer haze is slit by the runway. Moments before touchdown, a succession of images marks his descent—a rooftop becomes a pylon to wheel the plane around base leg to final approach. Runway numbers freeze in the windshield, then pull the plane down an imaginary wire to touchdown.

At night these visual references vanish. Huge tracts of landscape run to the horizon and disappear. The edge of an airport is blurred in the glimmer of a nearby town. Instead of sharp outline and perspective, the eye is tricked by illusion—a narrow runway seems longer, a wide field apparently grows shorter. Interstate highways ominously resemble lighted runways (Fig. 16) and invite the pilot to land. The final approach to the real runway may be a skittery slide over invisible obstacles.

The uncertainty and deceit disappear under the lights at night. Nearly every daylight cue continues to glow after dark through a dozen different lighting systems. Approaching a distant airport on a clear night, the pilot may see the glint of a beacon a half hour away. As he nears the field, runway edges gather between strings of white lamps. In poor visibility,

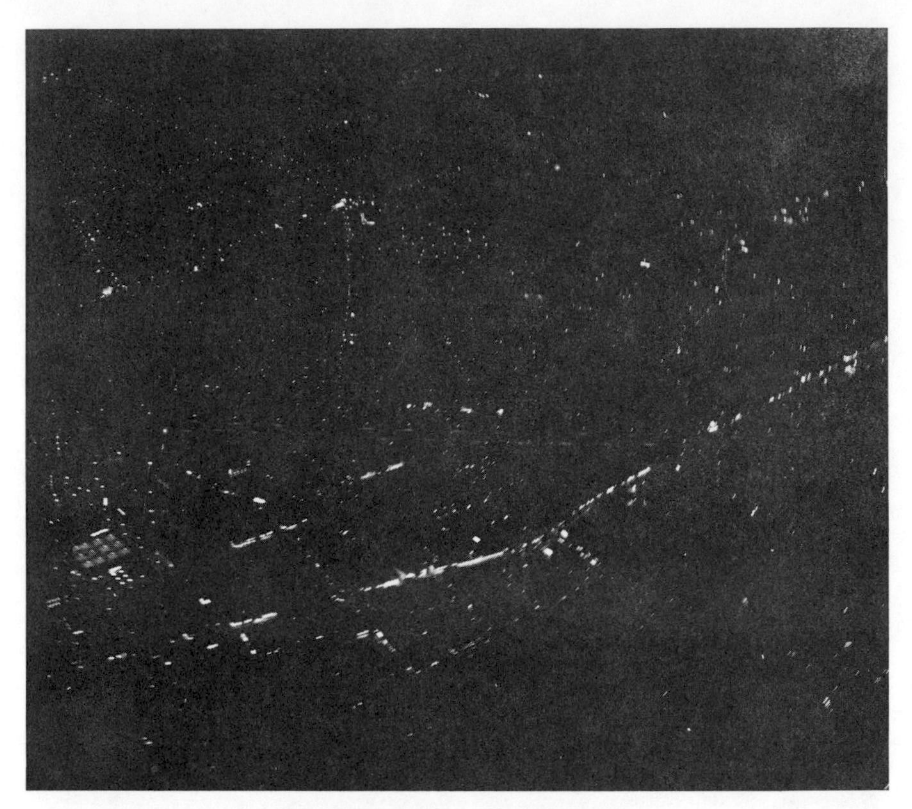

16. **Confusion of city lights.**

17. **Lighted wind tee has 40-watt green lamps atop crossbars.**

high-intensity strobes capture the eye and draw it to the runway threshold. Now the pilot glides safely down an optical path formed by slanting shafts of red and white light. Landing direction appears on the crossbars of a lighted wind tee (Fig. 17).

Lighting is an enormous boon to the night flier, but it also has its share of challenges. It often glitters in cryptic codes that need deciphering. Some facilities, like one at a small field in Ohio, date back to the scarf-and-goggle days. A pilot landing after dark must circle the field twice to rouse the airport manager out of bed. The manager turns on his car lights to indicate the direction of landing! At the other end of the spectrum is the international jetport. Approaching the edge of New York City, miles from J. F. Kennedy Airport, the pilot sees a brilliant line of lead-in lights. They form glowing tracks across the borough of Brooklyn, cross the border into Queens, then split into luminous forks on the final approach.

Most lighted airports are between these extremes. And here lies the challenge to the night flier. Will a landing be greeted by smudgepots, strobes, or an electrified ball of lightning that hurls itself against the runway? Perhaps the field will be utterly dark on arrival. (One small strip turns on its field illumination only if the pilot pays a small lighting fee.) Other airports, with excellent lighting systems, extinguish them after evening business hours. The lights snap on automatically—but only if a pilot knows how many times to press his microphone button. All these details are readily available before take-off.

A rich source of lighting information is the sectional aeronautical chart. It does not yield every detail, but strokes in the essentials. Glance along a route of flight; a star above an airport symbol instantly reveals a rotating beacon (Fig. 18). If an "L" appears in the airport information box (Fig. 19), some form of field lighting is on throughout the night.

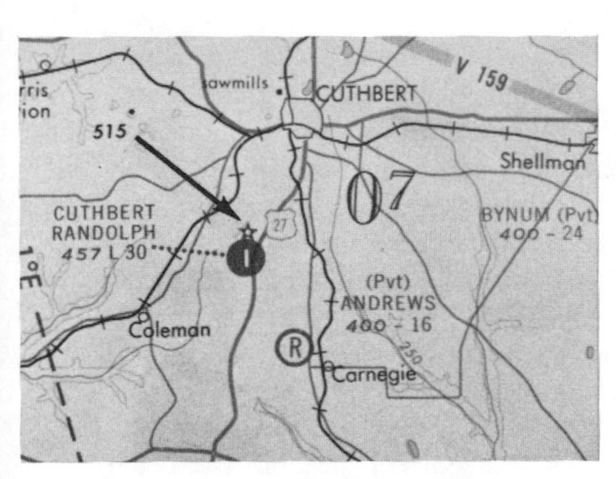

18. Star indicates rotating beacon.

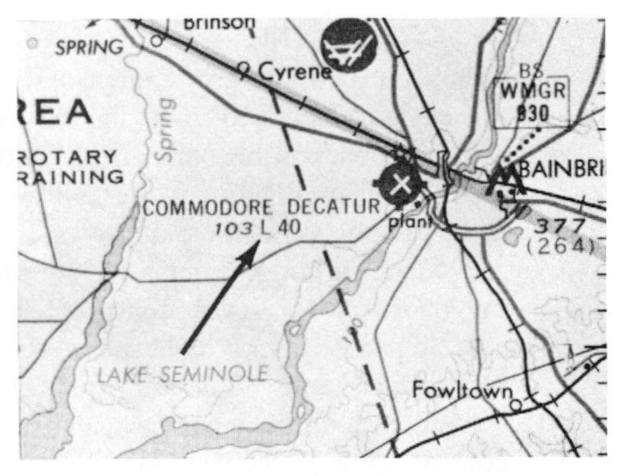

19. "L" indicates field lighting.

Knowing these facilities—a beacon and field lighting—is vital if an urgent situation occurs en route. Sudden loss of oil pressure, bad weather or some other critical development may demand an immediate landing at the nearest airport. A lighted field is selected in moments on the sectional. Since the chart is usually carried in the cockpit, it may be the pilot's single most valuable lighting reference.

Airports that don't shine their lights throughout the night are also indicated on sectional charts; if the letter "L" is surrounded by parentheses (L), be warned the field is illuminated only part of the night (Fig. 20). Another precaution is the asterisk: *L (Fig. 21). It means that lights are provided only when the airport is given prior notice. Requests can be by radio call, a written letter, or even a telegram.

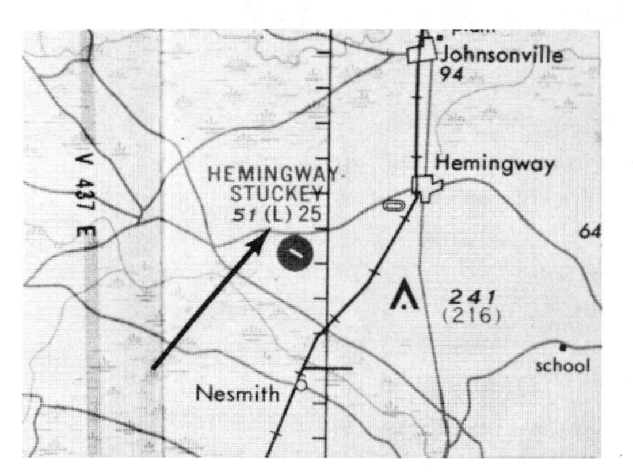

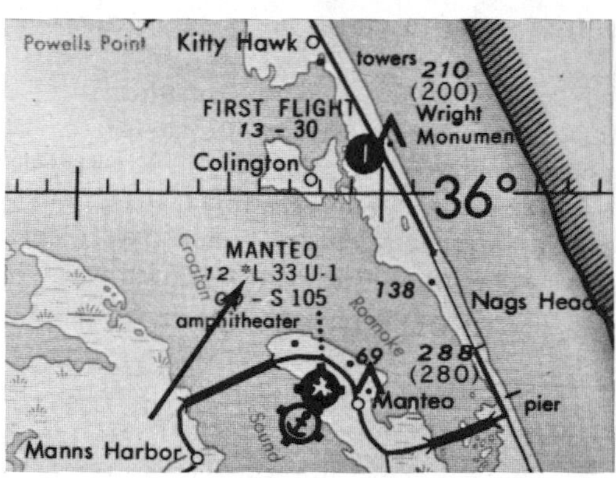

20. (L) means lights are on only part of the night.

21. *L requires prior call for lights.

So look to the sectional as the first key to lighting. Besides emergency value, the chart is an excellent aid in preflight planning. Nighttime routes are easily selected to favor lighted fields. When darkness hides landmarks, rotating beacons become visible checkpoints along the way. Another bit of lighting intelligence is the yellow area indicating a city or town. Populated regions glow in a pattern which resembles their outline on the chart.

Some lighting systems on a sectional are not aeronautical. The flashing or occulting lights of a marine buoy (Fig. 22), for example, are reassuring when flying over water, the most trackless topography on a dark night. The same buoy, however, can be misleading if it guards the entrance to a harbor that lies near an airport (a common combination for coastal cities). Unless the pilot plucks the correct beacon from the twinkling array below, he could descend into a hazardous area. In the ex-

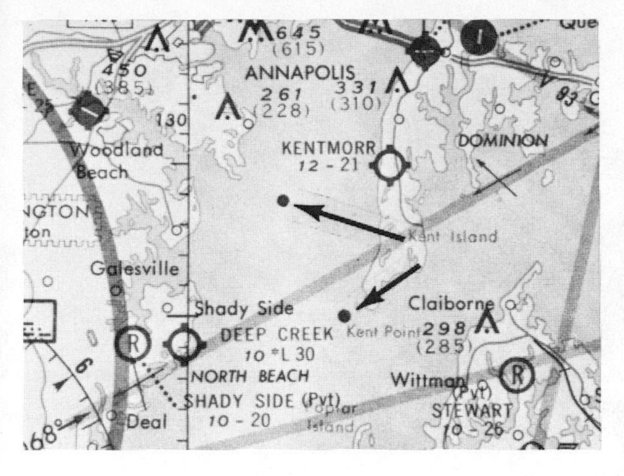

22. **Marine buoys.**

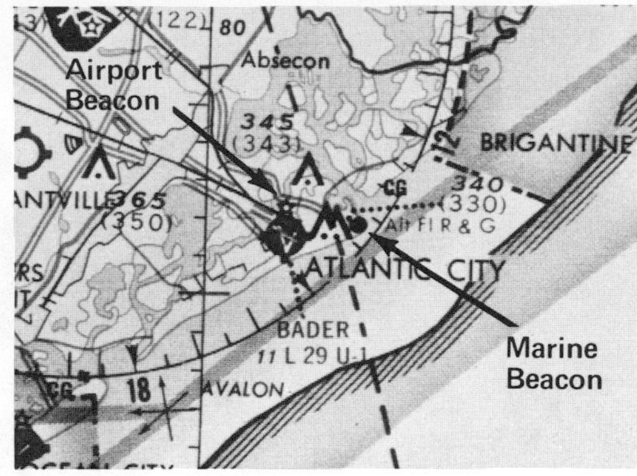

23. **Marine beacon next to airport.**

ample shown in Figure 23, a flashing red and green marine beacon is lo cated a scant two miles from the Atlantic City airport. Fortunately, the color and flash rate of a marine light hardly resemble those of an airport beacon. (Air navigation lights and their symbols are shown in Figure 24.)

AIR NAVIGATION LIGHTS

Rotating Light	☆
Rotating Light (With flashing code lights)	–.. ☆
Rotating Light (With course lights and site number)	12 ☆
Flashing Light	Fl ☆
Flashing Light (With code)	Fl ☆
Lightship	
Marine Light	Occ W R G ●
Rotating Light (On top of high structure)	1504 ☆

F—Fixed
Fl—Flashing
Qk Fl—Quick Flashing
I Qk Fl—Interrupted Quick Flashing
Occ—Occulting
Alt—Alternating

Gp—Group
R—Red
W—White
G—Green
B—Blue
(U)—Unwatched
SEC—Sector
sec—Second

Marine lights are white unless colors are indicated; alternating lights are red and white unless otherwise indicated

24. **Air navigation lights shown on chart.**

25. Airport rotating beacon. Two lamps, 180° apart, produce six white and six green flashes per minute.

Rotating beacon

If any one light should be committed firmly to memory, it's the airport rotating beacon—friendliest sight on a faraway horizon (Fig. 25). The first spark of white may attract the eye fifty miles away. To tell whether it is an airport beacon amid the myriad other lights, watch and wait. Confirmation comes five seconds after the white flash—a green light follows it around the sky. If green and white alternate twelve times per minute (or at five-second intervals), you can shift attention from charts, instruments and navigation. The arcs of light lead you to the airport. Safety is enhanced because you can make a head-up approach toward the beacon and watch for other aircraft converging on your destination.

Don't try to judge distance by a beacon's brightness. The powerful 36-inch rotating beacon at a large airport with high-intensity lighting has a 1,200-watt lamp. Smaller airports use a 10-inch beacon with a 620-watt lamp. (Sometimes the large-size beacon is installed at a small airport where there is high background brightness and possible confusion from neighboring lights.) The 36-inch beacons generally cast light two degrees above the horizon over level terrain, while small beacons are set for six degrees. The final angle is determined by flight tests. An aircraft is flown at an altitude commonly used by approaching traffic, and the beacon is adjusted until the pilot sees the center of the beam.

Airport beacons don't mark an airport's geographic center, or even lie close to it in many cases. The beacon at a large airport, say FAA guidelines, should not be closer than 750 feet from the centerline (actual or extended) of the nearest runway. At a small airport (with a maximum runway of 3,200 feet) the beacon is held at least 350 feet away from the main runway's centerline. In most locations, expect a beacon to be no further than 5,000 feet—nearly a mile—from the nearest usable landing area.

An exception occurs in hilly country where the light of a beacon is shielded from arriving aircraft. Now the beacon may be moved up to two miles away from a runway and mounted atop high terrain. When this is done, an extra light—called an identification beacon—is installed on the airport near a runway. If the airport is on land, an identification beacon flashes green. Flashing yellow is for water airports and flashing white for unlighted airports.

The most common rotating beacon, by far, is the one atop a civilian land airport. Driven at six rpm, green and white lenses emit light every five seconds. If the white light glints with a *double* flash, the field is a *military* airport. A landing here is costly unless you have prior permission or it's an emergency. The aeronautical beacon for a lighted *water* airport is an alternating white and *yellow* flash. (Again, the white is given a double peak to warn that it's a military installation.) White flashes *only* indicate an unlighted airport, but this is rare. Heliport beacons flash in a green-yellow-white sequence. Aeronautical beacons are turned on when the sun's disk is more than six degrees below the horizon —equivalent to sunset and sunrise—and some are illuminated during the day when ceiling and visibility are restricted.

Obstruction lights

As you fly toward an airport at night, other beacons may appear. Red flashes warn of obstructions. Alternate white and red flashes are emitted from *landmark* beacons (Fig. 26) atop prominent features of the terrain or high man-made structures.

Since collision with obstacles near an airport is a serious cause of light-plane accidents, the FAA has jurisdiction over structures that may intrude into the airspace. When new construction is planned near an airport, the builder notifies the FAA about lighting such hazards as electric poles, wires, silos, fuel tanks, amusement-park rides and raised roads.

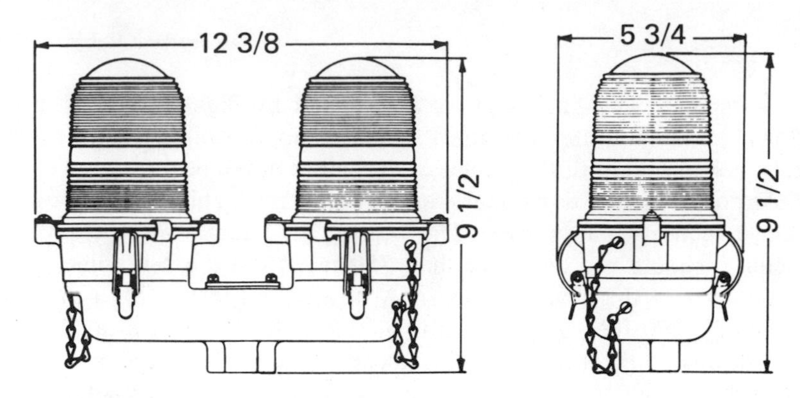

26. Obstruction marker lights. Double units are used atop narrow obstructions or at edges of extended obstructions, with single units spaced equally between them.

Near large airports the FAA may allow a structure to rise one foot for every hundred feet of spacing from the nearest runway, while small airports are allowed a one-foot elevation for every fifty feet the structure is removed from the runway. These ratios keep the glide path reasonably free of obstacles around the public-use airports listed in the AIM (Airman's Information Manual) directory. Two hundred feet above ground is the general dividing line to determine whether obstacles (any distance from an airport) need an obstruction light.

The tremendous altitudes reached by TV broadcast antennas, transmission wires and skyscrapers exceed the warning capacity of a flashing red lamp. This has led to more efficient systems based on the high-intensity strobe. A tall tower may be fitted with three strobes—top, bottom and center—with flashing rates of about sixty per minute in a middle-top-bottom sequence. In one installation across the Mississippi River near New Orleans, bright strobes delineate 460-foot-high transmission towers which bear the cable. The lights do not mark the cable, but the supporting towers. (Early attempts at attaching warning lamps directly to cables touched off power failures because of wind vibration.) The lights are reduced in brightness at night (from 200,000 candelas to 1,000) and tests in the early 1970s showed strong pilot approval. One flier said he could see the strobe lights while looking directly into the setting sun.

Runway lights

An airport beacon flashes the location of a field, obstruction lights keep you high and wide of obstacles, and runway lights guide you over the final few feet between air and earth. Runway lights are not intended to illuminate the surface. Some light spills onto the runway, but their pur-

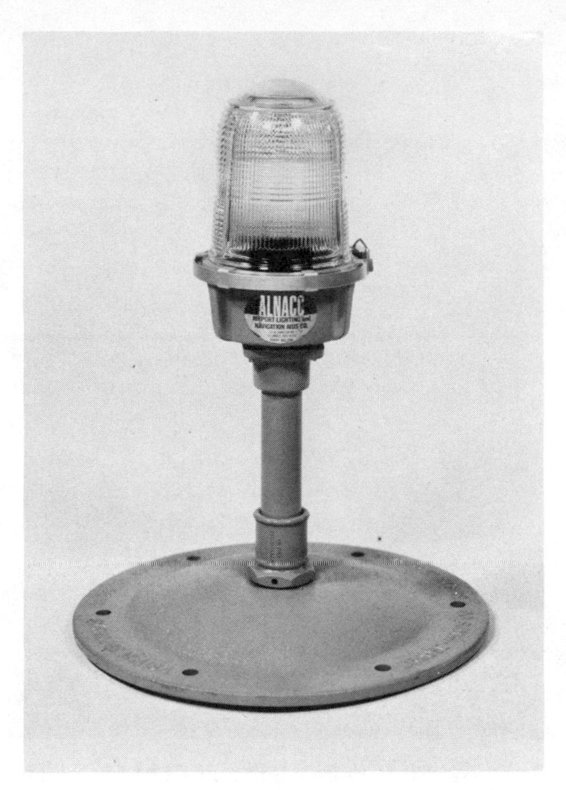

27. Medium-intensity runway light to mark runway and threshold.

pose is to outline left and right runway edges (Fig. 28). Seen from altitude, they also help a pilot align the plane along downwind, base and final legs.

The simplest runway lights, the Low Intensity type, are popularly called "flying farmer" lights. They are inexpensive, easy to install and convert a sod strip to night operation. The bulbs are only 15 watts, which limits them to good weather and sparsely populated regions where few competing lights can overwhelm the runway outline.

Medium Intensity is the next step up for runway lighting. The lamps are 40 watts for greater visibility, but higher power is only part of the improvement. Each lamp is fitted with a lens to lift the illumination into the approach path to give the pilot a better view of the runway at greater distances. These focused beams radiate in opposite directions so the view improves on an approach from either end. The lights are also aimed so that the pilot isn't dazzled as he arrives over the threshold.

High Intensity lights are the deluxe illumination of larger airports. Since they complement an ILS (Instrument Landing System), the lamps are 200 watts to cut through low visibility. These lights have powerful beams to guide an aircraft during an instrument approach, but they won't blind an arriving VFR pilot. When weather is good, the tower reduces

28. Runway lights define left and right edges of field.

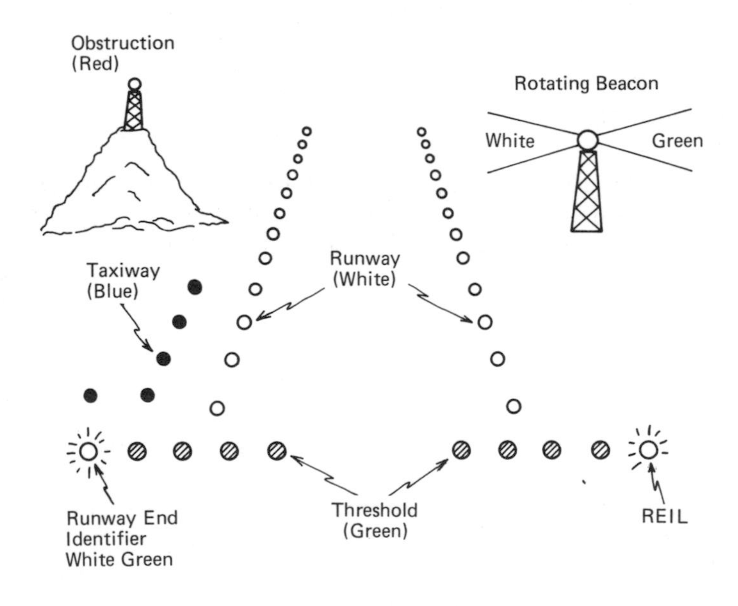

29. Basic airport lighting.

High Intensity lights far below their maximum brightness. They can be adjusted to any of five brightness levels by radio request from a pilot. (Lost pilots have occasionally picked out a runway from a confusion of city lights by asking the tower: "Flash your runway lights, please.") To caution the pilot, the last 2,000 feet of runway lights are yellow, and red lights cross the runway's end.

REIL (Runway End Identifier) Lights are an excellent guide to a runway because they present a clear, easily understood image. They are a pair of high-intensity strobes, straddling the runway threshold, which flash twice a second (see basic airport lighting in Fig. 29). When you are miles from the airport the REIL's piercing twinkle resembles no other lights on the terrain. Closer in, they also furnish visual guidance to the runway by radiating compressed beams. Always fly toward the *less* intense lamp; when both lights appear equal you are on an extended centerline of the runway. The high brightness of REIL lights won't blind you as you come over the threshold, because the beams are tilted away from the runway centerline.

Threshold

Most airports don't have the boon of REIL lights, and a runway threshold is marked only by three or four green lights at right angles to the runway. Know the green threshold lights well—and never touch down before them! In many cases these lights do not mark the physical beginning of a runway. If wires or a weak surface present a landing hazard, the runway threshold is *displaced* to keep aircraft from touching down in an unsafe area.

The part of a runway closed for landing by a displaced threshold is often suitable for taxiing and take-off. To prevent waste of valuable real estate, color-coding the runway lights confers certain privileges inside a displaced threshold. Lenses in these lamps are masked by filters to show the following colors, depending on the light's position and your taxiing direction.

No lights preceding the threshold. Remain out of this area for any operation. You may be tempted to land or taxi here because the runway surface might gleam in the moonlight. The absence of lights edging this area, however, means a ditch or other hazard to landing, take-off and taxiing.

Red lights. When white runway lights change to red, the colors communicate two instructions. Do not *land* in the red area. You may, however,

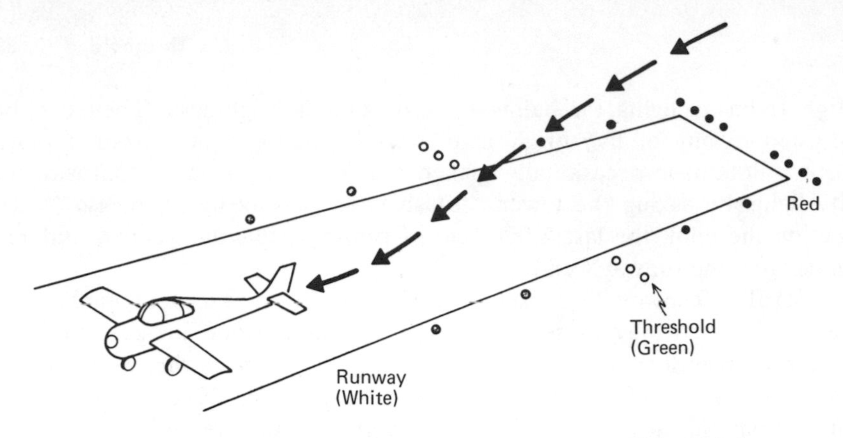

(A) Do not touch down in area bordered by red lights (right).

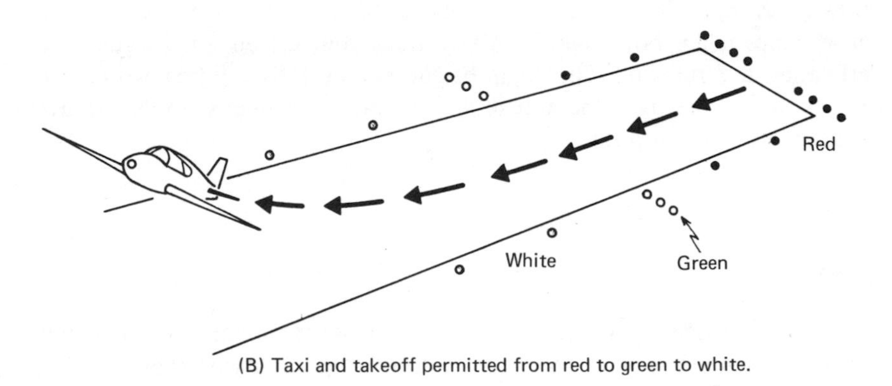

(B) Taxi and takeoff permitted from red to green to white.

30. Displaced threshold lighting.

taxi and *start* a take-off in the red area if green threshold lights appear *ahead*. In other words, you can use this section of runway for taxiing and take-off, but not for landing.

Blue lights. These mean taxiing only. Never take off between blue lights. This may be confusing since known runways are sometimes bordered by blue lights, but they are in service as taxiways.

VASI

The VASI system is a tremendous boon to night flying. Meaning Visual Approach Slope Indicator, it is an array of lights near the runway threshold which raises narrow beams along the approach path. The beams appear to have different colors, depending on whether the aircraft is above, below or on the correct glide path. In the common two-color VASI (Fig. 31) the pilot sees only red lights if he is too low and only

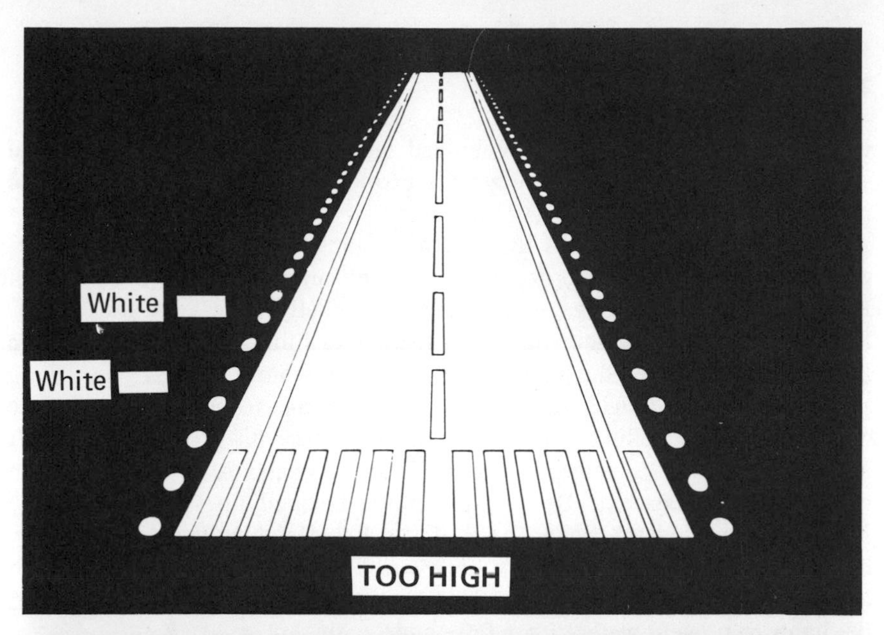

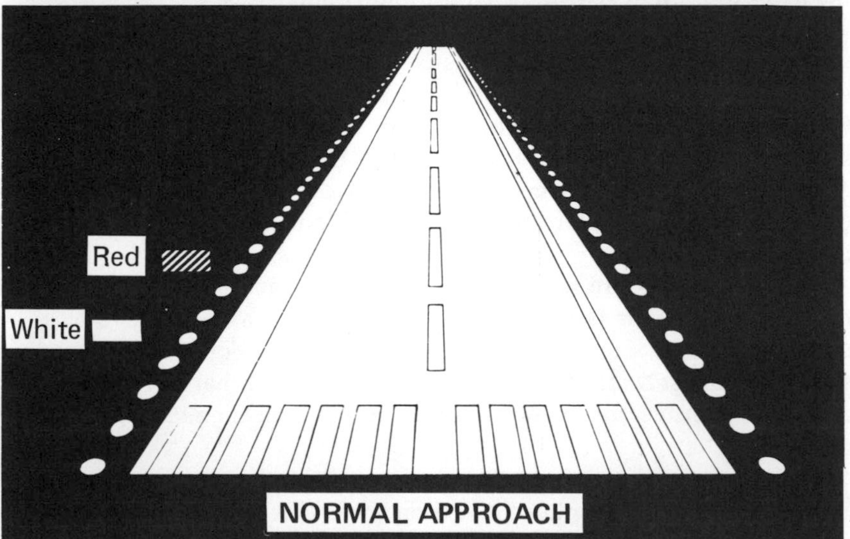

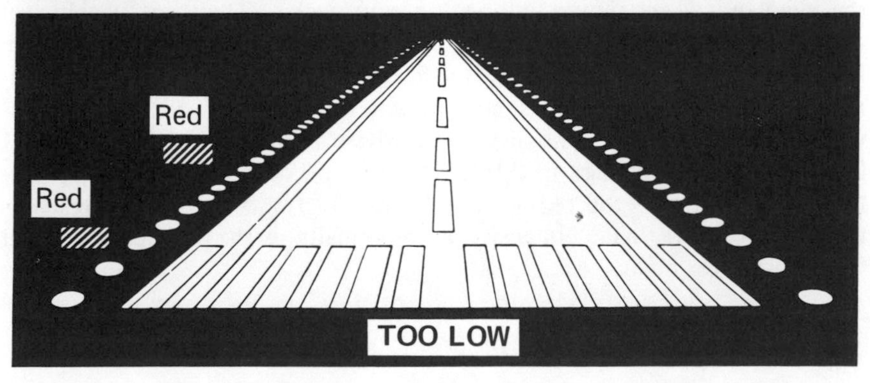

31. Visual Approach Slope Indicator (VASI).

white lights if he is too high. When red and white lights appear in equal strength, the plane is descending on a proper glide path that ends at the runway threshold.

The path of a VASI is inclined at an angle between 2.5 and 4 degrees from the horizontal, depending on the airport. If the field has an ILS, the visual path agrees with the electronically generated glide slope. Flying down a VASI demands no instruments other than the pilot's eye and his ability to keep the plane centered on the beam.

A white VASI light appears yellow or orange-to-brown under some conditions of air pollution and darkness, but these disturbances shouldn't affect your ability to interpret the red-white signals. A precaution when using the VASI system: Be sure the plane is lined up with the runway *before* interpreting the lights. Also, if the lights suddenly extinguish during an approach, something may have struck the ground fixtures. In such cases, a tilt switch built into a VASI assembly automatically turns off the lights to prevent generating a false glide path.

VASI lights eliminate the danger of a straight-in approach to an airport at night. As described in a later chapter, a pilot's depth perception is poor at night unless he flies a normal traffic pattern to relate his altitude and position to the runway. With a VASI for vertical guidance, however, the confusion of random ground lights or the flat void of water are removed. From distances of four miles or more from the end of the runway, a pilot can intercept the visual glide slope and descend straight in with confidence.

VASI systems are not all alike, but the variations aren't great. Some have a total of four boxes (Fig. 32), while large airports have as many as twelve boxes, six on each side, as shown in (Fig. 33). Touchdown should occur between the downwind and upwind bars. These systems are too expensive for small airports, so more economical VASIs have evolved.

SAVASI. Meaning Simplified Abbreviated Visual Approach Slope Indicator, this system has two boxes. The units are placed only at the left side of the runway (when viewed during the approach). They are usually set about fifty feet from the side of the runway, level with the runway crown. When it's not practical to locate the boxes at the left because of topography or economy, the lights may mount on the right side of the runway. The FAA requires the airport to advise pilots of a right-side VASI by publishing the change in AIM. SAVASI needs no special interpretation because it uses the red-white code of the full VASI. Some installations control the lights' intensity automatically with photocells: bright for day, dim for night.

32. Approach to a VASI with four light boxes (two near each side of the runway threshold).

33. Close-up of one side of twelve-light VASI system.

A three-color VASI system combines all signals in a single box. Instead of red and white beams, the device sends slanting yellow, red and green beams to the approaching pilot. Yellow warns that the plane is too high, red means it's too low. When the pilot is on the glide slope he sees a green shaft that is usually projected at an angle of two degrees to create the desired glide path.

Approach lighting systems

Lighting described so far is mostly designed for VFR operation. The systems furnish all the visual information a pilot needs to locate an airport, glide toward the runway, flare, roll-out and taxi. The instrument pilot, however, traces much the same path through solid cloud via electronic references displayed in the cockpit. In the final moments before landing, electronic signals may not be sufficiently accurate to guide the plane to touchdown. During the transition from IFR to visual contact with the runway the pilot is guided by the ALS, or Approach Lighting System.

A single-engine pilot should know how to interpret the ALS, even on cloudless nights. It forms a perplexing array that could easily mislead an unwary flier. As the name implies, ALS mainly illuminates the *approach,* not the runway, area. The light groupings form apparently flat regions which invite a touchdown when, in reality, the lights are mounted on poles or towers thousands of feet before the runway threshold. A premature descent on the ALS would be a disaster. As shown in Figure 35, some approach lights are mounted on piers over water, while others intrude into an adjoining town (Fig. 36).

Approach lights are installed in a variety of configurations. Their formats are not difficult to comprehend because the basic outlines are similar. Economy approach lights, for example, are not drastically different from those of an international airport, but have fewer lamps compressed into shorter distances.

MALS, or Medium Intensity Approach Light System, is in the economy class. Its layout, shown in Figure 37, helps a pilot align his craft with the approach path and keep the wings level. A standard MALS starts 1,400 feet in advance of the runway threshold with seven rows of light bars straddling an imaginary extended centerline of the runway. Each row is spaced 200 feet from the next one, but they apparently fuse together when viewed from a distance. The overall appearance is that of a solid swath of lights abruptly ending at the runway threshold. A pilot knows exactly where the runway begins—and where it's safe to touch down—by the green color of the threshold lighting.

34. Approach lights are mounted in advance of the runway threshold.

35. Approach lights mounted over water.

The second major feature of MALS are 1,000-foot markers, sometimes called "roll bars." They are extra light stations to the left and right of centerline bars 1,000 feet from the runway threshold. To an approaching pilot, the overall image is that of a lighted cross with its top end toward the pilot.

36. Approach lights (lower left) extending into built-up area.

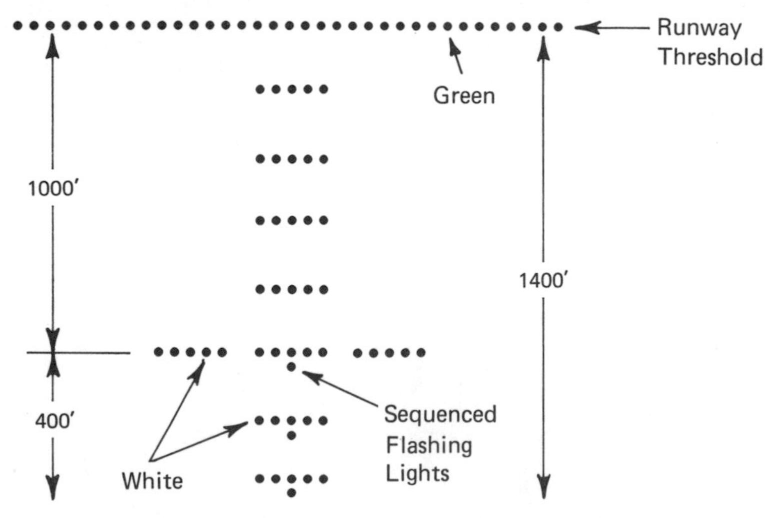

37. MALS (Medium Intensity Approach Light System) with sequenced flashers (strobes).

MALSF. The system is upgraded with the addition of SF, a Sequenced Flasher. It is a series of powerful strobes at the three outermost light stations which repeatedly flash from the furthest point inward. It has the visual impact of a lightning ball hurling itself against the runway. No other light is as distinctive or compelling to the eye. During an approach over a lighted city, the sequenced flasher quickly identifies the airport. Its primary task, however, is to penetrate cloud and fog in the final stages of an instrument landing. Popularly called the "rabbit" by professional pilots, it is usually the first light seen before breaking out of an overcast. The lights are normally off during good weather, but some controllers may turn them on to help a pilot find an airport in a confusing location. You may request a tower to do this, or to turn off the flashers while you approach under any weather condition.

MALSR. If any single lighting system can be called standard for many airports serving commercial carriers and heavy aircraft, it's this one. Hundreds were installed, starting in the early 1970s, to bring the benefits of approach lighting to small airports without the huge cost of earlier systems. It includes technological advances in strobes and pilot preferences not available before. A new element is the RAIL—Runway Alignment Identifier Lights, an improved version of the sequenced flasher, or strobe lights.

The layout of a MALSR (sometimes spoken as "MALS/RAILs") appears in Figure 38. Note that the standard length is 3,000 feet in advance of the runway threshold. The first 1,400 feet resemble the simpler MALS

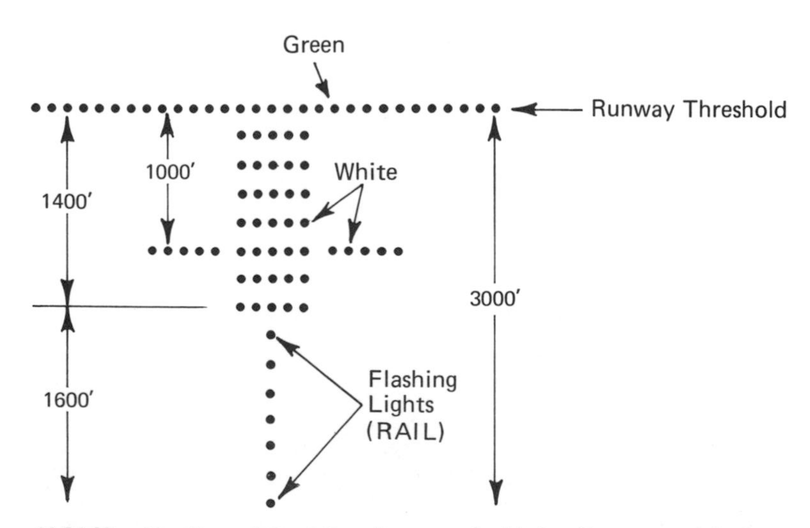

38. MALSR, Medium Intensity Approach Light System with Runway Alignment Indicator Lights (RAIL).

arrangement—that is, seven rows of bars forming a band of light extending 1,400 feet from the threshold and marked with extra horizontal bars at the 1,000-feet-to-go level. The innovation is a line of eight RAIL lights along the centerline for 1,600 feet in advance of the light bars. (All lights are high-intensity strobes that flash in sequence from the approach end toward the runway threshold. The illusion, as before, is that of an electrified rabbit moving at incredible speed toward the runway.

To avoid dazzling the pilot, RAIL lights have three brightness levels. A pilot may request the tower to switch RAILs to high, medium or low brightness. At the same time, the runway lights also vary in intensity. For reasons of economy (to avoid costly cables between the approach area and tower) RAILs are interconnected to the medium-intensity runway lighting, and both systems change brightness together.

ALSF-1 and -2. These early systems were installed at approximately 250 major airports and were still operational in the mid-1970s. As seen in Figure 40, ALSF-1 bears striking similarity to the approach lighting already described. An exception in ALSF-2 is the double row of red lights on either side of the light bars for the last 1,000 feet.

A simpler version, the ALSF-1, reduces the red lights to a single row just short of the runway threshold. This has become the standard for large airports with lower weather minimums. (For instrument pilots, MALS/RAILs are standard for Category I operations; while ALSF-2 is the required lighting system for Category II.)

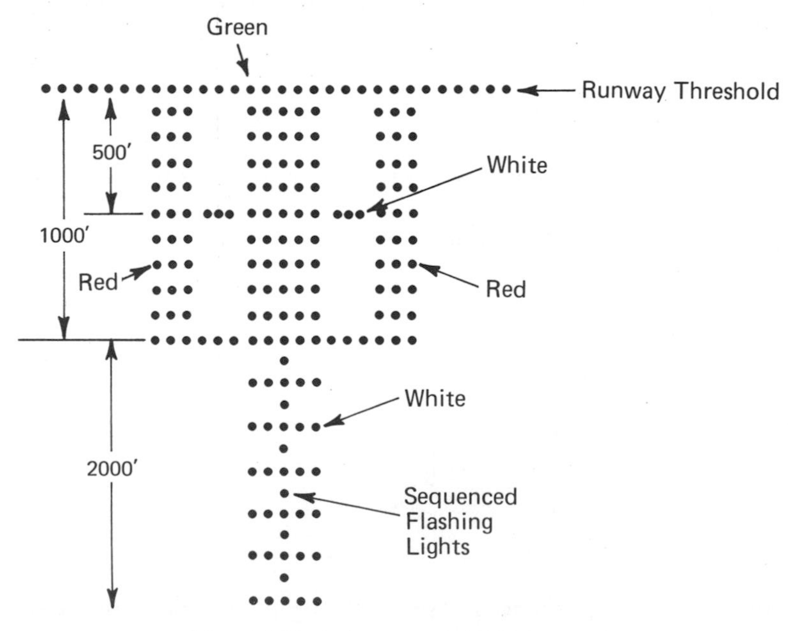

39. **ALSF-2 Approach Light System.**

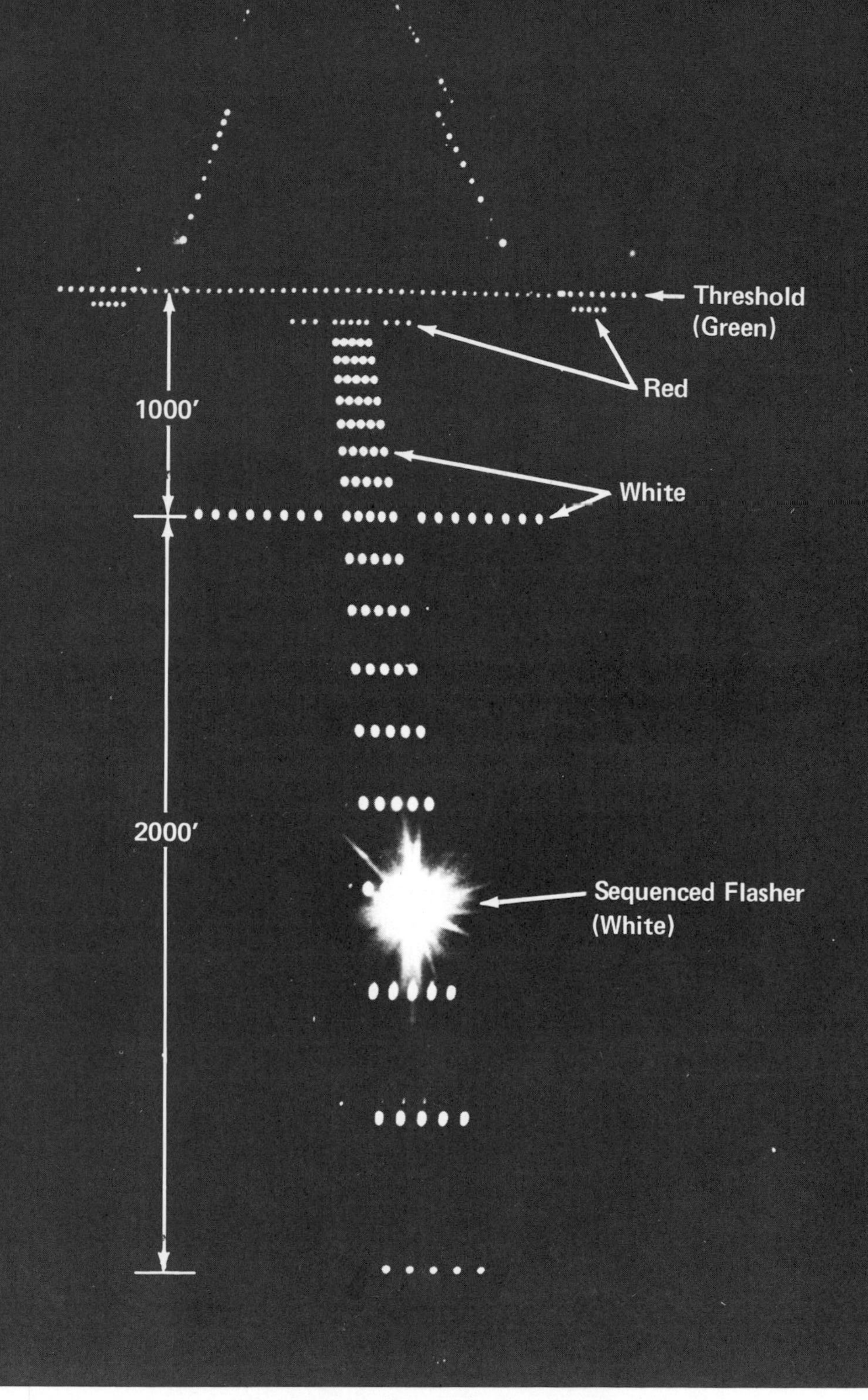

40. Actual ALSF-1 Approach Light System.

41. Centerline and touchdown zone lights.

Touchdown zone and centerline lights

At large airports, runway illumination is enhanced by additional lights mounted flush with the runway surface. Touchdown zone and centerline lights (Fig. 41) communicate valuable information to the airline or corporate pilot arriving over the runway threshold when forward visibility is cut as low as 700 feet. The lights give clues for flare, touchdown, roll-out and take-off.

Touchdown zone lights are primarily a landing aid. They form two broad ribbons of white light astride the centerline, from the threshold to 3,000 feet down the runway. (If the runway is less than 6,000 feet long, touchdown zone lights run only half the runway length.) By eliminating the so-called "black hole" appearance of a dark runway, the lights help a pilot start his flare and show how it's progressing.

Touchdown zone lights hold visual pitfalls for the VFR pilot. To someone unfamiliar with their purpose they appear as part of the approach lights (Fig. 42). Since most lights are mounted on posts, a pilot may hesitate to land directly on top of two illuminated bands. The illusion is heightened (as Fig. 43 shows) by a black runway which seemingly elevates the lights above the surface. Touchdown zone lights, however, are recessed into the runway surface and are no obstacle to the airplane.

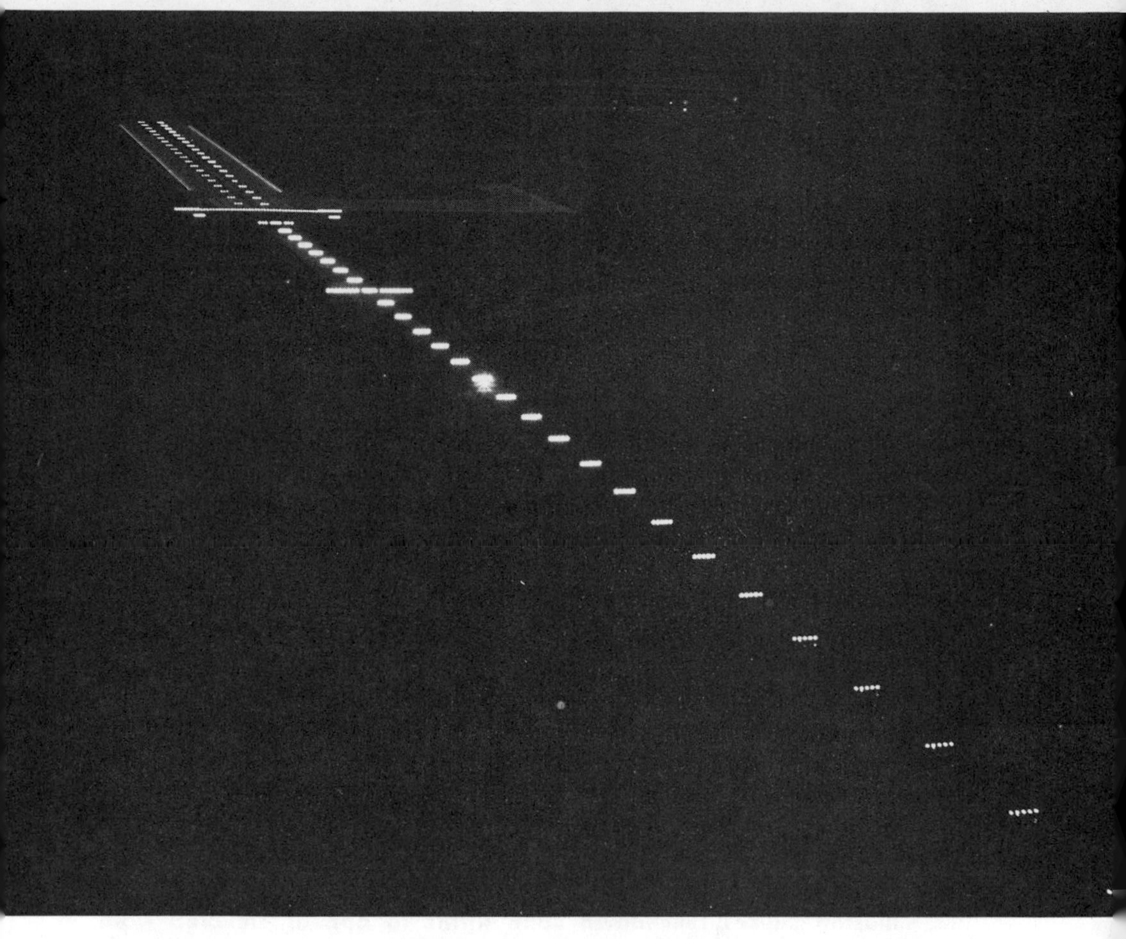

42. Touchdown zone lights, at upper left, should not be confused with approach lights, which stop at runway threshold.

A similar apparition appears when taxiing onto the runway for take-off. Touchdown zone lights seem to block your right-of-way, but they silently slip under the wheels.

Centerline lights replace the white-painted line seen during the day. Spaced 50 feet apart, they are slightly off-center (up to 2 feet) to avoid interfering with the painted stripes. White centerline lights run from the threshold until they reach the last 3,000 feet of runway. Then they alternate red and white for 2,000 feet. You'll know 1,000 feet of runway remains, since the centerline lights turn completely red in this area. These distance markers are more significant to the captain of a heavy aircraft, but should be familiar to a light-plane pilot. A small plane might be given an intersection take-off near an area of red centerline lights but still have ample room for take-off. A lighted centerline is also an excellent reference for keeping the airplane straight during a take-off or landing roll.

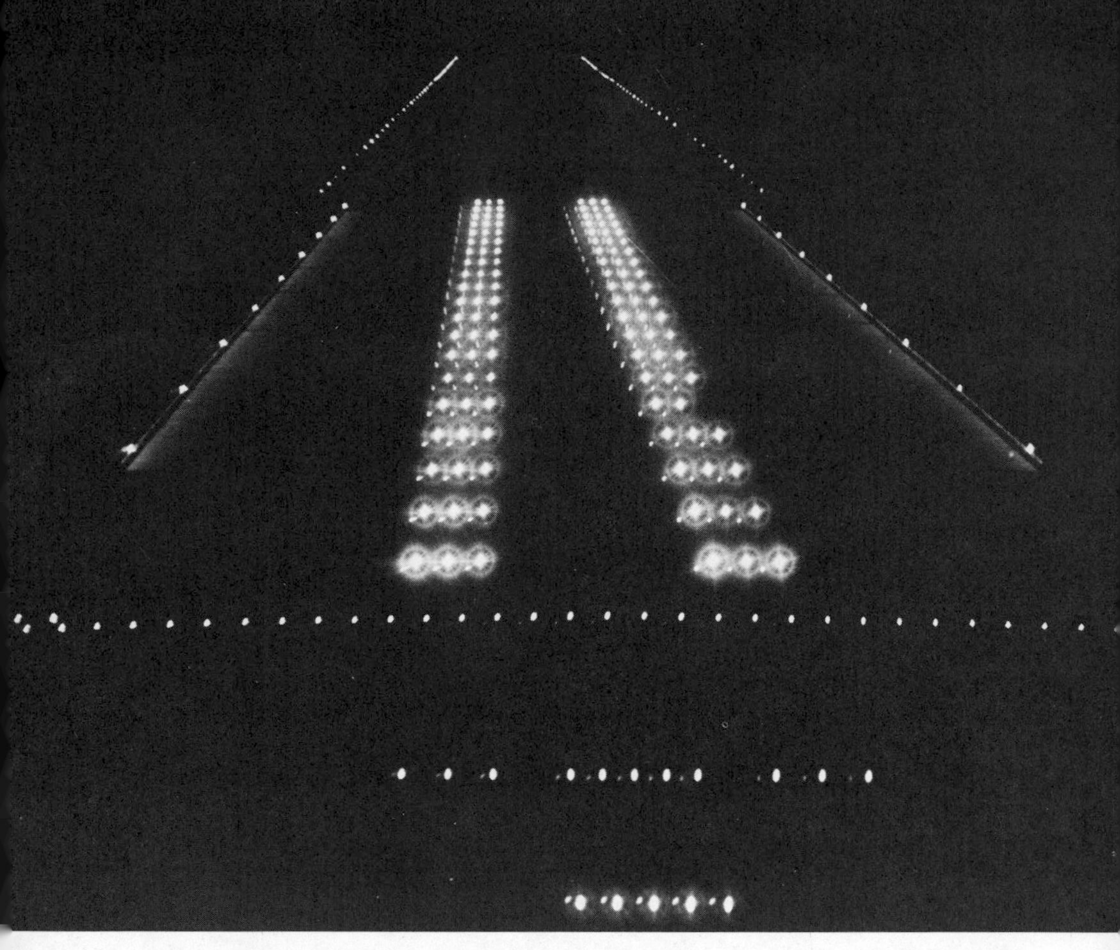

43. Illusion causes touchdown zone lights to appear elevated from runway surface.

Airport lighting is not only variable from one field to the next, but is a subject undergoing considerable research. There are experimental landing-direction pointers, new varieties of obstruction lighting and taxiway signs. The basic outlines and colors, however, tend to persist: green threshold, blue taxiway, white runway lights, the transverse white bars of an approach-light system with its 1,000-foot marker and abrupt termination at the runway threshold.

Lighting information

If you plan to navigate to an airport at night, much detail about its lighting system is available before the flight. Start by finding the field in Part 2 or Part 3 of AIM and check its lighting entry. It is deciphered with information shown in a Legend (Fig. 44) located in Part 2. Many small, lighted airports, for example, are in the "BL4" class. Expect a ro-

LIGHTING

B: Rotating Light (Rotating beacon). (Green and white, split-beam and other types.) Omission of **B** indicates rotating light is either not available or not operating standard hours (sunset-sunrise).

L: Field Lighting. An asterisk (*) preceding an element indicates that it operates on prior request only (by phone call, telegram or letter). Where the asterisk is not shown, the lights are in operation or available sunset to sunrise or by request (radio call). **L** by itself indicates temporary lighting, such as flares, smudge pots, lanterns.

1—Portable runway lights (electrical)
2—Airport Boundary
3—Runway Floods
4—Low Intensity Runway
5—Medium Intensity Runway
6—High Intensity Runway
7—Instrument Approach (neon)
7A—Medium Intensity Approach Lights (MALS)
8A, B, or C—High Intensity Instrument Approach (ALS)
9—Sequence Flashing Lights (SFL)
10—Visual Approach Slope Indicator (VASI)
11—Runway end identifier lights (threshold strobe (REIL)
12—Short approach light systems (SALS)
13—Runway alignment lights (RAIL)
14—Runway centerline
15—Touchdown zone

Because the obstructions on virtually all lighted fields are lighted, obstruction lights have not been included in the codification.

44. Legend in AIM for deciphering lighting details.

tating beacon and a field illuminated by Low Intensity runway lights. It may be assumed these lights remain on from sunset to sunrise. If "B*L4" appears, the asterisk is a warning that prior request must be made to have lights turned on. How the request is made appears in a footnote. The footnote may also prohibit night landings on certain runways or in given directions (Fig. 45).

An instrument-rated pilot has another excellent source of lighting information; the airport diagram on instrument approach plates. The exact site of the rotating beacon is given with the familiar star (see Fig. 46A).

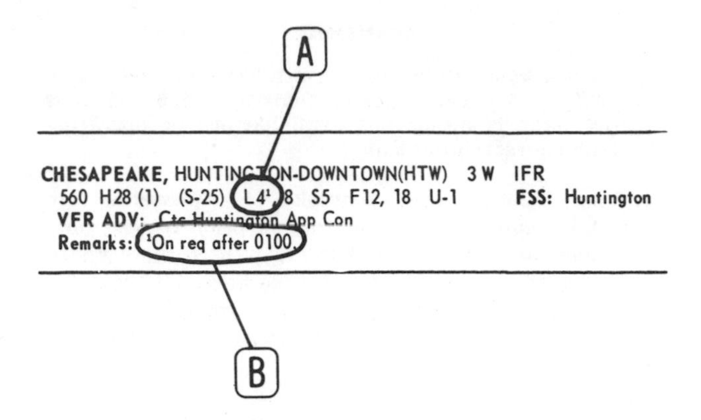

45. For this airport, AIM directory shows lighting facilities as "L4[1]," meaning Low Intensity runway lights (see A). Footnote B warns that prior request must be made for lights after 0100.

If a rotating beacon doesn't have a standard flash rate, a dot-and-dash code next to the star indicates how it flashes. Lighted obstructions usually appear on these charts when they're close to the flight path, but you can expect a lighted field for public use to have its threatening projections topped by red lighting. Consider Dulles International Airport (Fig. 46B) as an example of how elaborate lighting may become. It has centerline lights down two runways, several "V" symbols which indicate "VASI," and runways tagged with "A" symbols for approach lights.

Radio control

An intriguing system adopted by many airports is lighting by radio control. Instead of burning lights continuously during the night, these airports have a special monitor receiver (Fig. 47) which responds to signals from an approaching aircraft. If the signals follow a specific code, the lights are automatically turned on. This is attractive for small airports because it saves a considerable amount of power (and labor costs for renewing burned-out lamps).

The "code" which triggers these systems can be transmitted by an airplane with a two-way radio. The pilot merely presses his mike button in a certain sequence. After the lights come on, they glow long enough (about fifteen minutes) for a complete approach and landing. If something delays the landing and the runway threatens to darken at a critical moment, a pilot keys his mike again and is granted additional time peri-

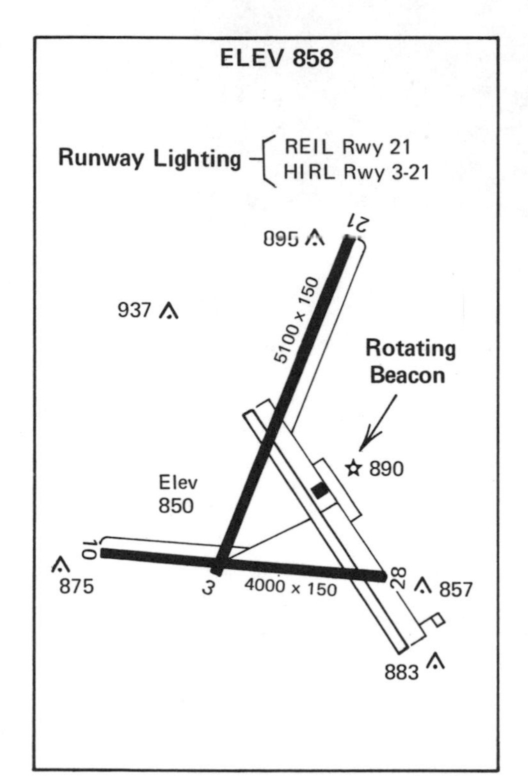

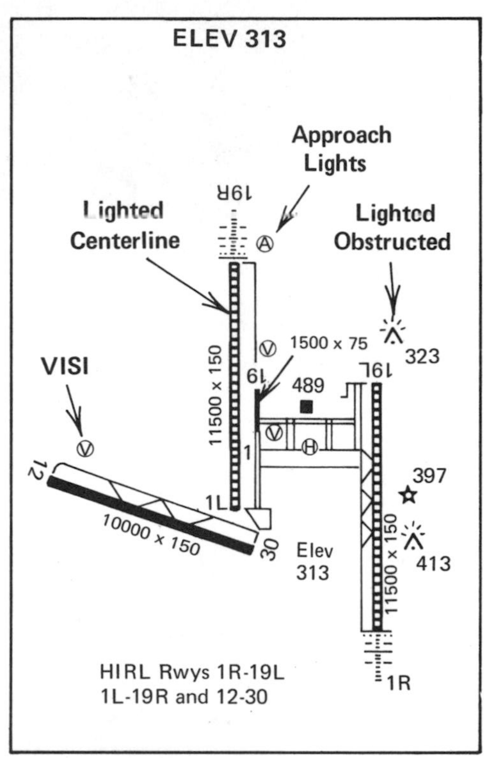

46. Lighting information shown on instrument approach charts.

47. Special airport receiver automatically turns on runway lights by radio control.

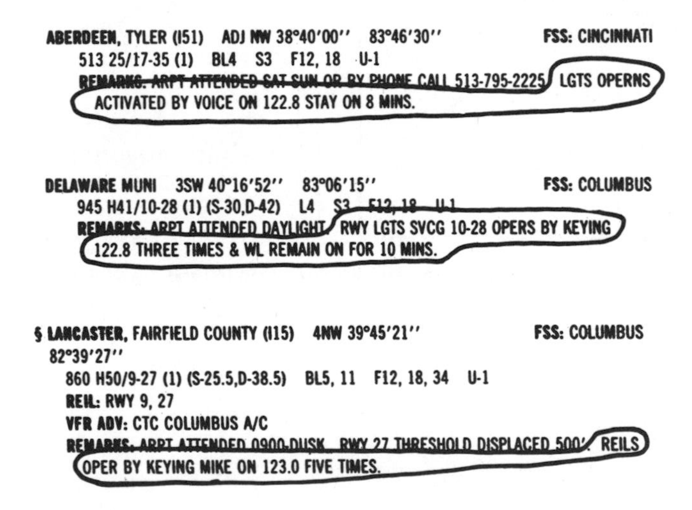

ABERDEEN, TYLER (I51) ADJ NW 38°40'00" 83°46'30" **FSS:** CINCINNATI
513 25/17-35 (1) BL4 S3 F12, 18 U-1
REMARKS: ARPT ATTENDED SAT SUN OR BY PHONE CALL 513-795-2225. LGTS OPERNS ACTIVATED BY VOICE ON 122.8 STAY ON 8 MINS.

DELAWARE MUNI 3SW 40°16'52" 83°06'15" **FSS:** COLUMBUS
945 H41/10-28 (1) (S-30,D-42) L4 S3 F12, 18 U-1
REMARKS: ARPT ATTENDED DAYLIGHT. RWY LGTS SVCG 10-28 OPERS BY KEYING 122.8 THREE TIMES & WL REMAIN ON FOR 10 MINS.

§ **LANCASTER**, FAIRFIELD COUNTY (I15) 4NW 39°45'21" **FSS:** COLUMBUS
82°39'27"
860 H50/9-27 (1) (S-25.5,D-38.5) BL5, 11 F12, 18, 34 U-1
REIL: RWY 9, 27
VFR ADV: CTC COLUMBUS A/C
REMARKS: ARPT ATTENDED 0900-DUSK. RWY 27 THRESHOLD DISPLACED 500'. REILS OPER BY KEYING MIKE ON 123.0 FIVE TIMES.

48. "Remarks" entry in AIM directory shows three different types of lights by radio control.

ods. In another arrangement the airport runs its lights at only 10 per cent brightness, then automatically boosts them to 100 per cent on a command signal from the approaching plane. Still another variation is radio-controlled REIL lights. REILs are kept off because of expensive, short-lived strobe lamps, and their brilliance may annoy nearby residents if allowed to flash continuously all night.

In planning a night flight, you want to know about a radio-control system at your destination and how to trigger it. Some older systems simply react to a human voice; you talk into the mike and say, "Lights, please." Since lightning discharges and other interference may falsely trip these lights, voice-triggered systems are largely supplanted by mike keying. A common code is five presses of the mike button within five seconds (with no voice). It doesn't require clocklike precision to make it work.

Besides knowing how to key (or speak into) the mike, you must place your aircraft radio on the correct frequency. A popular channel is the unicom frequency (122.8) of uncontrolled airports. Other radio-control frequencies include 122.7 and 122.9 (normally used for air-to-air); 123.0 (unicom at a controlled field); and 123.3 (used by gliders). These channels are not primarily for radio control, but the Federal Communications Commission permits such use so long as there is no interference with regular operation. Congestion on the popular unicom channel (122.8) spurred development of these other frequencies.

How do you obtain details on lighting an airport by radio? The airport operator informs the FAA to have pilot instructions (the number of mike presses and frequency, for example) published in AIM. As shown in the example (Fig. 48), it appears in the directory under the "Remarks" heading. You may find some airports with sophisticated setups, like one in Virginia, which lists two lighting frequencies for the same runway. One channel switches on flashing strobe lights to identify the runway's east end; the other channel illuminates lights at the west end. Thus you can light up your choice of landing runway, depending on wind direction.

Vision and vertigo

4

An instructor of aerobatics challenged his ground school students: "What would happen if you held a bowl of soup on your lap while I flew a plane through a loop?" The fledgling pilots weren't duped by the question and several students called out the correct answer: "Nothing." They knew that centrifugal force developed in the loop presses the soup toward the bottom of the bowl. The liquid would obey an artificial gravity.

Night flight invokes a similar conspiracy of gravity and centrifugal force. It is eloquently described by Byron Moore, an American Airlines captain, as he recalled the early days of flight in *The First Five Million Miles*. He tells how old-timers pushed through soupy weather at night to catch the faint glow of a beacon—"a tiny oasis of light in the black void." Hoping the weather would improve, the pilot pointed his wingtip at the light and circled. By the second and third circle, the light rose higher until it was directly overhead. The fourth circle was usually fatal because the pilot unknowingly banked his wings to the vertical and the plane dropped out of control. Centrifugal force easily outwits man's sensory response to gravity.

Moore also describes an airmail pilot who lost altitude because of structural icing. As the plane fell to within 400 feet of the ground, the pilot desperately launched a flare. Instead of illuminating the ground the flare fell *up!* What should the pilot believe, his internal sense of where the ground lies or the immutable law of gravity that pulls a flare down? After a torturous decision the pilot picked the right answer. He defied feeling and flipped the plane upside down until its wheels were above the "ascending" flare. An instant later the plane struck the ground—right side up—which saved the pilot's life. What had happened was that loss of visual references had drawn the plane into a steep bank that rolled past the vertical. When the pilot saw the flare falling the "wrong" way, the airplane was nearly upside down. Veteran airmail pilots and jet captains are prone to disorientation when visual props are removed. Flying by the seat of the pants simply won't work.

That's why night flying has moments of instrument flight, where the panel must supply visual references hidden in the dark. During a routine maneuver—a turn, an ascent—the plane enters an attitude that eradicates the horizon. It is not, however, IFR flight. Most elements of instrument flying are missing; the separation between aircraft, flying in clouds, precision navigation to a runway or low ceilings and visibility. When fog or clouds lower toward VFR minimums, a prudent night flier stays on the ground. But in good weather he can go aloft safely if he can occasionally check and correct the airplane's attitude when the sky outside blackens everything from sight. The first time it happens is probably before the sun disappears below the horizon.

Taking off into the sunset

Night flights often begin with a roll toward the setting sun. If the disk touches the end of the runway, its glare may wipe out wires, hills and obstructions. It is one of the first visually stressing situations of night flight, but it's easily avoided by scanning the panel just after lift-off. An artificial horizon or turn needle assures a wings-level climb, as other instruments verify the ascent. Your peripheral view of the runway and terrain can supply the directional reference in the face of the sun's glare. When landing with a setting sun *behind* you, be aware that you may be invisible to other traffic about to take off.

Flicker vertigo is another disturbing condition that occurs when the sun is low. If the glowing disk is viewed through a propeller spinning between four and twenty times per second, the light has a stroboscopic effect not unlike the electronic flashes of a psychedelic light show. In some individuals it causes nausea or even serious disorientation. If you feel discomfort, *don't stare* through the spinning blades. Varying rpm with small throttle changes also helps.

Flicker vertigo doesn't happen while accelerating down the runway, because fast-spinning blades are invisible to the eye. You may sense it, however, at low rpm while taxiing on the ramp or rolling into position on the runway. Those critical rpm rates also occur on a final approach to landing into the sun. Looking away from the prop should quickly relieve any feeling of distress. Flicker vertigo is also triggered by the reflections of a rotating beacon or strobe after a plane has penetrated a cloud. These lights must be turned off during these conditions.

Night vision

It is widely known that the human eye cannot instantly adapt to the dark. This is demonstrated on entering a movie theater; aisles, seats and people merge in the blackness. After a while, outlines and detail appear as the eye is sensitized to the feeble illumination. The pupil dilates and nerve endings in the retina—the rods—become chemically charged for night vision. Although rods are nearly color blind, they operate with remarkable sensitivity in low light levels. About a half hour of darkness fully activates the rods, but a single flash of a bright light can desensitize the cells and ruin night vision. (Daytime vision is accomplished by another set of retinal cells called cones. They are sensitive to color and small detail.)

The easy loss of night adaptation has provoked extraordinary steps to preserve it. There's the classic scene of combat pilots in a ready room wearing red goggles. Since rods are relatively insensitive to red light, the pilots suffer negligible loss of night vision. The principle also appears in the internal lighting of aircraft. You may look at an instrument panel bathed in red light, then look out the window and see ground objects under an average moonlit sky without loss of night vision.

The 1970s started a trend away from meticulous protection of night vision. Airliners converted from red to blue-white instrument lighting, and many general aviation manufacturers followed suit a few years later. The reason is that most visual references shifted from outside to inside the cockpit. The era when a contact pilot had to grope in the dark for an unlit landing strip is over. Instead of squinting for a tiny beacon of light, a pilot glances at needles in front of him. On arrival, the airport is rimmed with bright white lights that attract the eye. Dr. Harold Brown, an authority on aeromedicine, remarks: "Problems such as improper fuel selection because of inadequate lighting are much more significant now than the loss of night vision."

If you want to keep your eyes most sensitive to the night anyway, follow this general rule: Use the least possible illumination. Reduce cabin lighting with the rheostat control to the lowest level. When you must look at a bright light, first close one eye to keep it adapted to the dark. Use this trick while taxiing near flashing strobes, beacons or landing lights.

Using a red lens or red cover on a flashlight while examining charts or dark corners of the cockpit is standard procedure, but since it is difficult to see some items under a red light, use a white beam and the one-eye-closed approach. Sectional charts have magenta-colored symbols which cannot be read under red light. Instrument-chart symbols are printed in purple or black ink and can be read in red or white light.

A brilliant flash of lightning or the intense lights of a city may also affect night vision. In unavoidable instances, turn up instrument lights to full strength until the disturbance has passed. Researchers have found that it speeds the transition back to night adaptation.

Studies by the U. S. Air Force also show that visual sensitivity in the dark is affected by a pilot's exposure to bright light during the day. Under extreme intensities (from snow or sand), night-vision loss may run as high as 30 to 50 per cent for many hours. Protecting the eyes against strong light by wearing sunglasses during the day is the remedy. Be certain to remove sunglasses at night before take-off.

Another enemy of night vision is altitude. The retina is one of the first tissues in the human body to suffer from a lack of oxygen (hypoxia). Ten thousand feet is the usual limit recommended for light-plane flying with no oxygen, but night vision suffers at lower altitudes. At about 4,000 feet, there is a loss of 5 per cent of night vision; at 16,000 feet, the drop is measured at 40 per cent. The instrument panel is increasingly difficult to read in dim light and some loss of contrast occurs between objects. If you fly over 5,000 feet at night, you may have to turn instrument lights to higher illumination and accept less visual acuity when looking outside.

Illusions

A welter of optical illusions fools the healthiest eye after dark. The so-called "blind spot" is created by a scarcity of rods in the central portion of the retina. Gaze directly at a dimly lit object and you may not see it! Look slightly to the side of an area you wish to perceive at night to force it into view. This technique, called "eccentric vision," is mastered after a few trials.

"Focus fixation," another visual deception, occurs when few outside images reach the eye. The lens tends to focus at short distances and reduces your awareness of distant aircraft or obstacles. It's avoided by never staring at night and keeping the eye in frequent motion.

The "autokinetic" effect is a remarkable night illusion. It begins if you intently gaze at a single light in the distance. The object should re-

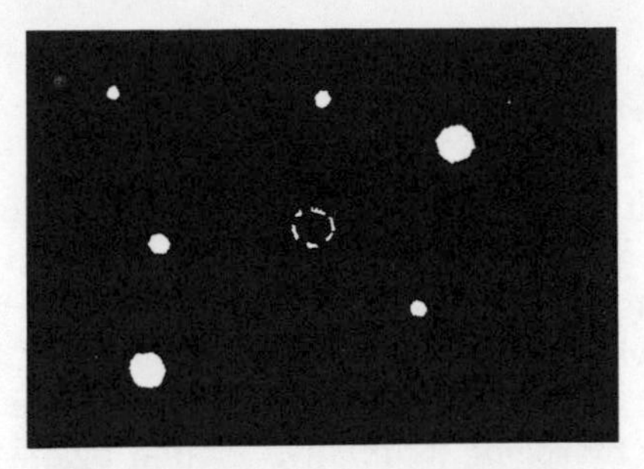

50. Blind spot. Staring directly at distant light causes it to disappear.

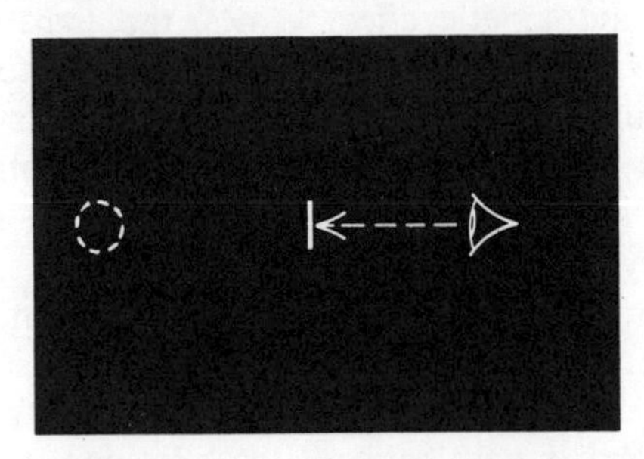

51. Focus fixation. Staring into dark causes short-range focus and loss of distant image.

52. Autokinesis. Staring at light causes false motion.

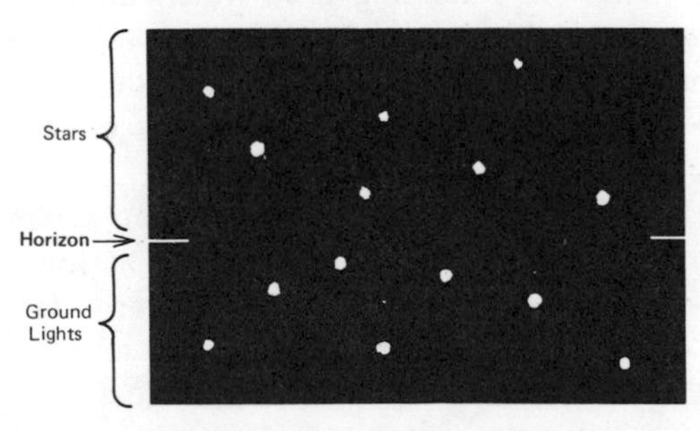

53. When no horizon is visible, ground lights may appear as stars.

main in place, but it starts to drift in unpredictable directions. I recall a case of autokinesis after the launch of the huge Echo I satellite. Newspaper reports said its progress would be visible to the naked eye at night as a bright star moving slowly across the sky. At the appropriate time I gazed in the correct direction—and every star became Echo I! The true satellite was verified by seeing its relative motion against fixed stars in the sky.

The illusion has been cited as a cause in military aircraft accidents. While formation flying over a dark sea, a pilot follows the light of the lead plane. As the autokinetic effect grows, the light apparently moves away from the plane and the following pilot trails it into a steep dive. For a light-plane pilot autokinesis is troublesome on a starry night. Stare at a single star and you may convince yourself it is another airplane. The remedy is to avoid staring, and to check for relative movement against other lights in the sky.

Poor depth perception seems to cause many night accidents attributed to "landing short of the runway." Judging distance is partly accomplished by the binocular arrangement of the two eyes, but this facility doesn't extend much beyond the airplane windshield. Distant depth is perceived by cues of relative size and position, intensity of color and shadow. Since most of these are obliterated at night, the eye refers to distant lights, indistinct silhouettes or masses of gray. All are poor depth references. Is a single light in a black sky the powerful landing beam of an aircraft twenty miles away? Or is it the position light atop the tail of an aircraft two miles ahead? Both may shine with the same intensity and size.

When utter darkness hides the horizon, scattered ground lights may look exactly like stars. Imagine the consequence of using these lights as a visual reference in a steep turn. The plane may end up inverted.

Altitude is also difficult to judge at night. When runway lights are dim, the illusion is that you're higher. Holding the plane in a nose-high attitude while looking at lights below also simulates higher altitude to the eye. Haze in the air does the same because it blurs ground objects and creates an illusion of distance.

The "black hole" effect is a landing area which appears bottomless because of a region's general bleakness. It's also created when a runway surface is dark in contrast to strong nearby illumination. Sometimes runway lights appear to dance in the air around the edges of a chasm where the runway should be.

Size offers little help. How can you decide whether you're looking at a short runway from high altitude, or a long runway at low altitude? In researching landing-approach phenomena, A. Howard Hasbrook developed this graphic comparison: Assume you are at an altitude of 210 feet approaching an 8,000-foot runway that's 3,200 feet in the distance.

54. "Black hole" effect. Dark runway (near center) appears bottomless.

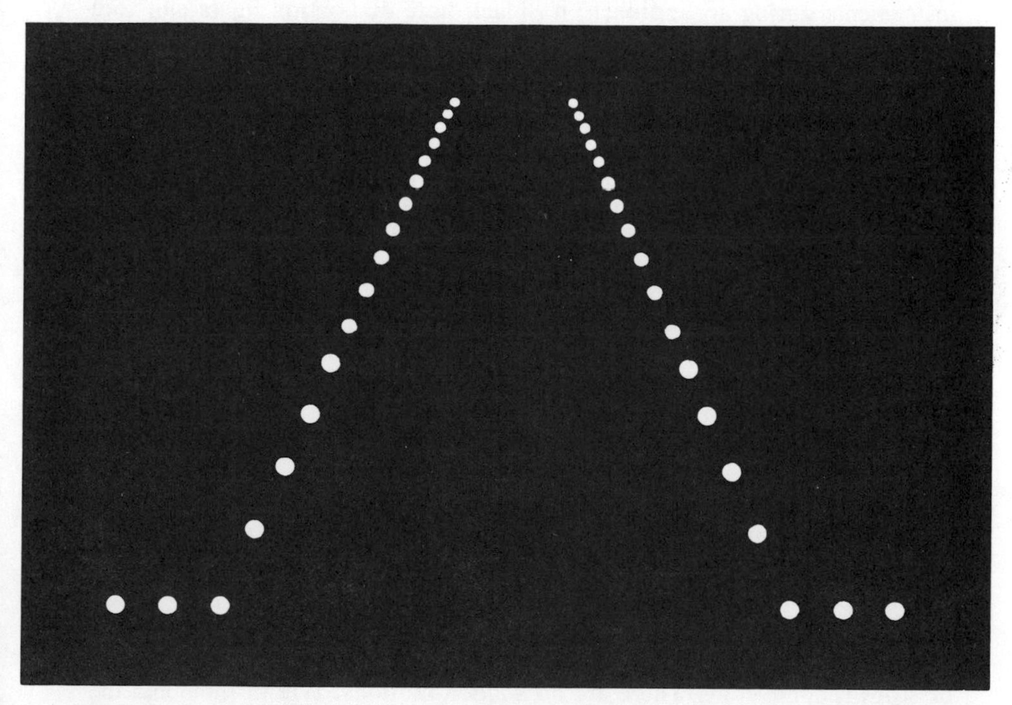

55. Runway shown above may be a short runway seen from high altitude, or a long runway seen from low altitude.

That identical image appears if the runway is only half that length (4,000 feet), from an altitude of 105 feet and a distance of 1,600 feet. As you can see in the illustration (Fig. 55), the eye receives no telltale hint of any difference.

A sloping runway also fools the eye. Since the dark withholds visible signs, it may be hazardous to land in mountainous country where runways commonly pitch up or down. When this warning was published in *FAA Aviation News* it drew an indignant response from a reader in a state that boasts few mountains. Kansas, he wrote, has at least *two* sloping runways. He went on to describe a personal experience in another presumably flat state, Georgia:

> I elected to land at Gunn Airport after making three passes down the dimly lighted runway. I didn't realize the runway sloped until I tried to set down and found myself still airborne with half the runway behind the airplane. A go-around didn't seem advisable because of a large hill half-obscured by the darkness at the end of the runway. You can believe I now memorize the "Remarks" in the Airport Directory.

His last remark is the greatest weapon against the illusions of night. Know in advance runway elevation and length, traffic-pattern altitude, and height and location of obstructions around the airport. Check your instruments during an approach; no black hole or floating lights can fool an altimeter. Know safe altitudes to fly and crosscheck against altimeter and vertical-speed indicator during the critical moments before landing.

Illusions are greatly reduced if you fly a normal traffic pattern around the airport at correct altitude. Straight-in visual approaches should be avoided unless you have the guidance of a VASI. If you have access to instrument charts for the airport, follow the altitudes recommended for the approach. AIM and sectional charts also provide important details. Keep scanning outside and integrate what you see with what is indicated on the instrument panel. Pilots who fixate only on the far end of the landing runway at night may forget to flare; if they gaze only over the nose they tend to "drop it in."

Vertigo

A spectacular moment of an FAA accident-prevention clinic happens during the vertigo demonstration. An experienced instrument pilot in the audience is asked to volunteer for a ride in a special chair. He is blindfolded and asked to "fly" the chair, much as he would an airplane. An FAA man gently rotates the chair (see Fig. 56), changes its direction or stops it completely. There are no sudden motions. Within moments the pilot is fighting his controls through violent maneuvers. When the blind-

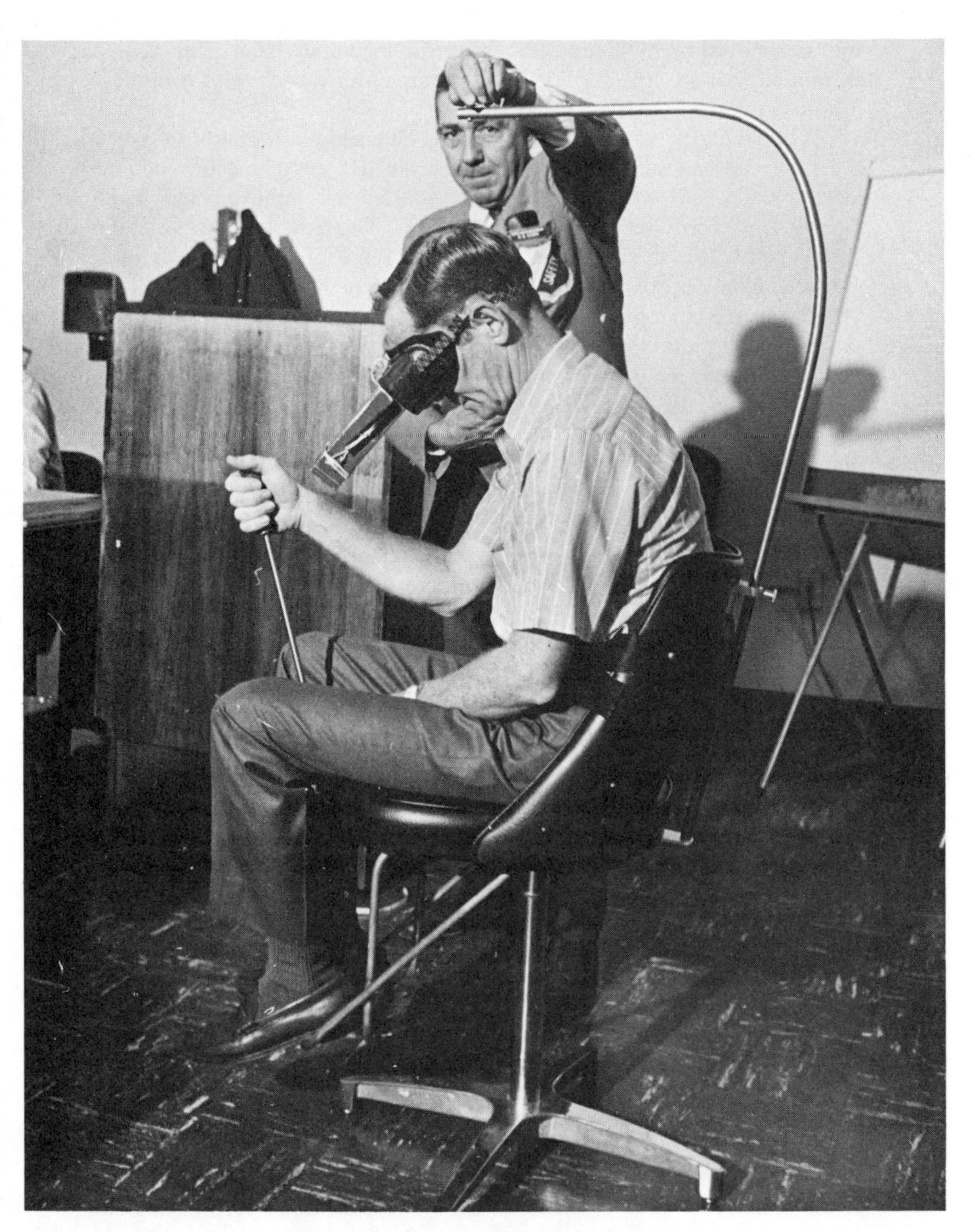

56. FAA vertigo demonstration.

fold is removed he blinks in disbelief at the inanimate chair. The audience is amazed that so little motion could communicate such wild attitudes to a trained pilot.

The volunteer was suffering vertigo. Not simple dizziness or nausea; the pilot had thousands of hours in the air. He experienced spatial disorientation caused when outside visual stimuli are removed and equilibrium is sensed only through internal feelings. The trouble is, those feelings cannot tell where the earth is. The human sense of orientation is mostly garnered from an eye which views the ground or horizon. When deprived of outside stimuli on a dark night, the pilot reverts to two remaining senses. One is the vestibular organ—a gyro-like structure in the inner ear. The other is the proprioceptive, or seat-of-the-pants, feeling of pressure on muscles and tendons. As the aerobatic soup bowl proved earlier, this fly-by-feel reaction is readily deceived as when, in a turn, centrifugal force supplants gravity.

The inner ear is an extremely sensitive indicator. Walk across a room with eyes closed and your progress should be as nearly true as when your eyes are open. As you move, three semicircular canals in each ear sense any position error and signal the brain to make corrections through the muscles. As seen in Figure 57, the canals function in any dimension because they lie at right angles to each other. They generate signals through moving fluids which stimulate nerve endings within each canal.

The semicircular canals are sensitive to changes in the body's *angular* motion; a companion structure responds to *linear* motion, or acceleration and deceleration. It's done by a sac of tiny granules which presses against nerve endings in the *static organ* (Fig. 58) as the body moves. Scientists believe this elegant sensory equipment evolved eons ago to accommodate one of the most fundamental human characteristics: the ability to walk upright.

But the delicate sensory equipment is linked to earthwalking, not airplane flying. The forces of flight easily fool the inner ear into sending the wrong directional signals. Because of inertia, inner-ear fluids cannot detect very slight changes of an airplane's attitude and fail to sense a gentle turn. Too, they are unable to perceive attitude changes if they occur at a constant rate. Even if a pilot suddenly realizes his airplane's attitude is wrong, his problem isn't over. If he tries to recover without seeing the horizon, inner-ear fluids spill beyond their rest position and tell the pilot he is now entering a similar maneuver, but in the opposite direction!

Such anomalies quickly appear during FAA demonstrations of vertigo. If the blindfolded pilot is asked to move his head downward as the chair turns, he confidently reports the airplane is in a roll, not a turn. He was fooled because his bending head changed the horizontal position of certain semicircular canals to the vertical. One fighter pilot during a chair

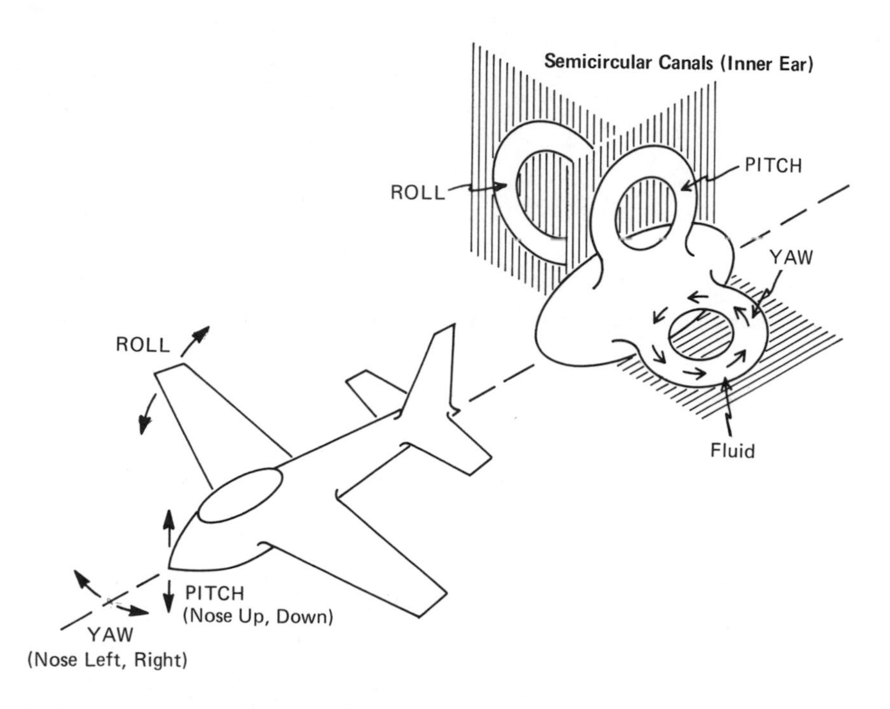

57. Fluid in inner ear canals signals motion. Based on normal gravity and level position, the organ is prone to error in flight.

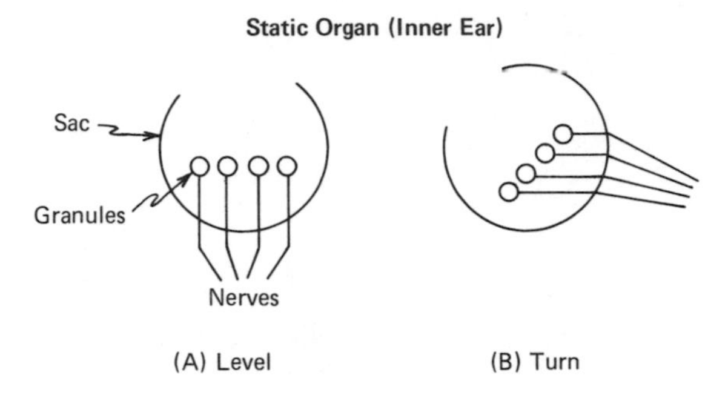

58. Granules bend under forces in turn and signal head position.

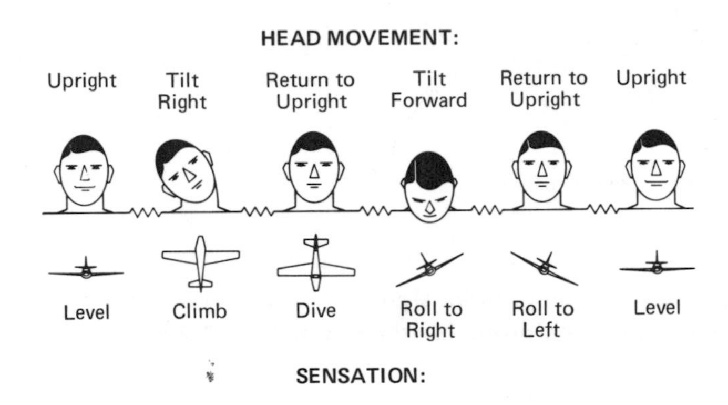

59. Pilot with vision restricted is seated in chair turning at constant 15 rpm. When ordered to move his head, as shown, he experiences false feelings shown on line marked "Sensation." Night flight can produce same conditions.

demonstration felt he was in the most "violent maneuver" he'd ever experienced. It was done only with gentle chair rotation and head tilting. Typical responses recorded by the FAA Office of Aviation Medicine are shown in Figure 59.

Flight instructors may give a student a dose of vertigo on a night checkout. It's easily started by putting the airplane in a turn, holding it there and telling the student to avoid looking outside. Next, the student is asked to make a sudden head movement. The spatial disorientation of vertigo quickly follows.

Vertigo is a killer but for the wrong reasons. Its confusion can be cured in moments by simple procedures. Catastrophe happens when a pilot believes his body's false signals and attempts to fly the airplane solely by feeling. Uncertainty develops into panic if the eyes catches sight of outside lights and tries to form a coherent picture. During this sensory struggle the plane escapes control and chances of recovery rapidly fade.

Vertigo can be prevented. If it's already established it's also easy to cure. The crucial rule: Whenever there are scant or unreliable visual references outside, never turn your head while the airplane is in a turn. Certain routine chores encourage vertigo: picking up a pencil, looking down at a chart, or switching a gas-selector lever. Avoid them all in turn. Wait until the plane is in straight flight.

When visibility is poor at night, avoid long spiraling turns or continuous circling maneuvers. They encourage vertigo as you attempt to orient yourself between outside and inside references. The two may not agree.

If your head starts swimming with the onset of vertigo, be aware that you can stop it quickly. First, keep your head erect. Closing your eyes briefly may help. Start scanning the gyro instruments to determine your attitude. Keep your eyes on the gauges until they indicate the plane is flying straight and level. Force yourself to ignore protesting sense organs and seat-of-the-pants feelings, because they may fabricate an entirely different flight attitude and direction. Believe the instruments and vertigo should pass.

Weather after sunset

5

About one third of all serious aircraft accidents are related to weather. At night the percentage rises sharply; estimates say half of all accidents after sunset are somehow involved with weather. In the classic daytime accident the pilot continues VFR flight into adverse weather, and the ensuing tragedy ends in one of two ways. After penetrating a cloud the pilot loses control because he can't muster the basic instrument skill to turn back to safety. In the other case he is flying too low and strikes a mountain or other solid obstruction.

The night pilot is already close to these provocative conditions. Visibility is reduced and the horizon an indefinite distant line. Now when weather drops toward visual minimums, IFR flight is nearly at hand. Further deterioration is serious unless a pilot can extricate himself on instruments.

The weather threat at night is not entirely bleak. Forecasting isn't exact but meteorologists' accuracy for nighttime predictions runs about 80 per cent. Those ominous statistics are slashed by a thorough preflight briefing, setting higher nighttime minimums, avoiding weather traps after dark, and maintaining basic instrument proficiency. About one in five nighttime weather accidents is aggravated by drinking.

It is distressing to read night accident reports because they reveal an appalling disregard of weather. One Cessna-172 pilot took off from Fort Lauderdale in perfect weather one afternoon and headed toward Washington, D.C. When he attempted to land at a Maryland airport at 10:04 P.M., poor visibility obstructed his forward view. Two miles from the runway the plane struck a pair of 230,000-volt high-tension wires and crashed. Miraculously, three occupants escaped with minor injuries. The altimeter in the wrecked aircraft was later found pinned to 700 MSL—thirteen feet below the power cables. In the investigation, the pilot declared he had obtained a weather briefing which called for VFR and improving conditions.

The FAA's report took a different turn. It read:

> Official U. S. Weather Bureau forecast for the period indicates a complex low pressure system moving from the Ohio Valley northeasterly at 20 knots. Northeast of this front in Maryland, Virginia, Delaware and D.C., expected ceiling 800/1500 feet broken, variable overcast at 4,000 feet, occasional overcast 200/600 feet, rain showers, fog tops to 18,000 feet, locally ceiling 200 feet, visibility one-half, rain showers, fog, chance of few isolated thunderstorms in the area.

The testimony of a witness, shown in Figure 61, verified the Weather Bureau's prediction.

Safety authorities consider basic VFR minimums—1,000-foot ceiling and 3-mile visibility—too liberal for night flight. In the "12 Golden Rules," a list of standards insurance companies believe could eliminate 93 per cent of all accidents, is this advice: "Never attempt a night flight unless you're sure you'll have a 2,000-foot ceiling and 5 miles visibility and will encounter no frontal fog, ground fog or storm conditions." Those minimums are nearly twice as high as basic VFR.

They are also far above "Special VFR," a type of clearance which ended for the night flier in 1972. In this operation a pilot obtained a clearance to fly within an airport control zone when visibility was a mile or better and he could remain clear of clouds. The advantage was that a private pilot without an instrument rating did not have to abandon a flight in generally good weather when marginal conditions were confined to local areas. In 1972, the FAA curtailed the SVFR rule because a survey showed the majority of twenty-three accidents connected with SVFR clearances happened at night. Controllers have stopped issuing those clearances after dark unless the pilot is instrument-rated and the aircraft equipped for IFR flight. At the same time the agency ended Special VFR for anyone flying in or out of several dozen major airports at any time. These terminals are surrounded by a circle of "T"s on aeronautical charts.

Marginal weather at night is rarely a total surprise, but ground fog is another matter. An astonishing prelude to ground fog is a marvelously

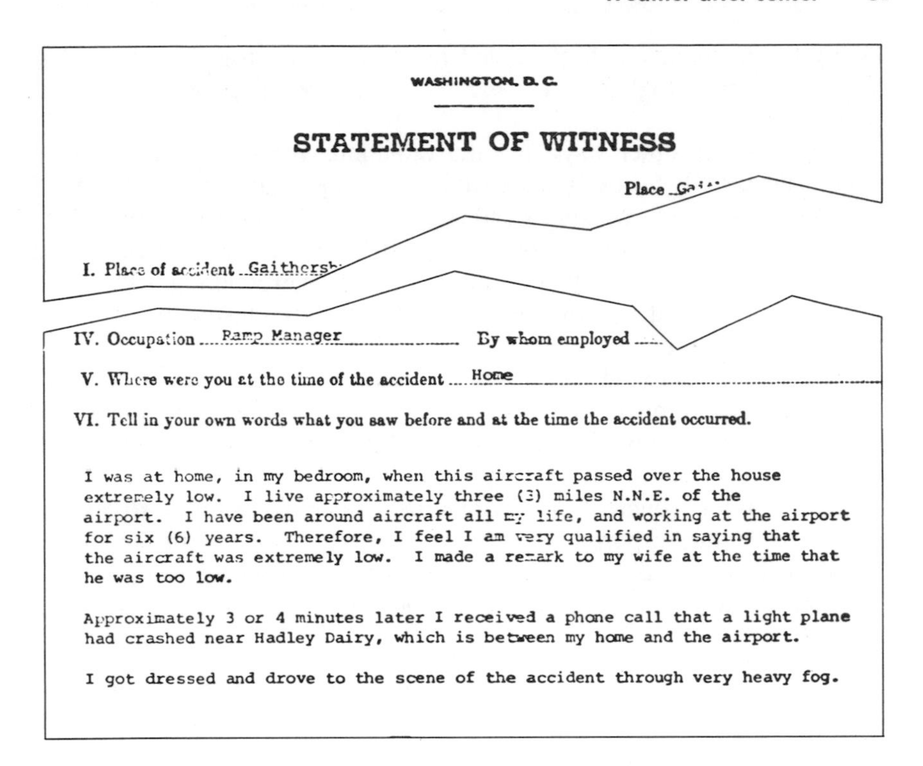
WASHINGTON, D. C.

STATEMENT OF WITNESS

Place Gai...

I. Place of accident Gaithersb...

IV. Occupation Ramp Manager By whom employed ...

V. Where were you at the time of the accident Home

VI. Tell in your own words what you saw before and at the time the accident occurred.

I was at home, in my bedroom, when this aircraft passed over the house extremely low. I live approximately three (3) miles N.N.E. of the airport. I have been around aircraft all my life, and working at the airport for six (6) years. Therefore, I feel I am very qualified in saying that the aircraft was extremely low. I made a remark to my wife at the time that he was too low.

Approximately 3 or 4 minutes later I received a phone call that a light plane had crashed near Hadley Dairy, which is between my home and the airport.

I got dressed and drove to the scene of the accident through very heavy fog.

61. Excerpt from accident report. Time was 10:20 p.m.

starry sky. The en route flight, too, is often unmarred by cloud or haze. On arrival at the destination, though, the descent encounters decreasing visibility. You may have suspected something amiss earlier, when ground lights grew dimmer, but as the plane nears the ground you are surprised to see the airport smothered in mist.

Ground fog is dangerous because it forms so quickly in otherwise excellent weather. The night is clear, the wind is calm. These conditions are part of the meteorological brew. After sunset the ground cools rapidly and chills the layer of air just above it. When the cool air reaches its dew-point, fog appears and tends to concentrate in flat areas. The height of ground fog varies with air currents, deepening under the mixing effect of a light wind. Thin fog, spawned in calm air, is still a serious hazard. It hardly appears threatening when you look down through it from altitude, but it can drop visibility suddenly to zero on an approach to landing.

Coastal regions are susceptible to ground fog. The Los Angeles basin floods with moist Pacific air that produces low-lying mist on quiet evenings. It collects in valleys as cool air tumbles down hillsides and douses the lowland in a steamy haze. A clear sky enhances the cooling effect. On an overcast night, clouds trap and reflect the earth's heat, preventing the temperature drop needed to form ground fog. Meteorologists call it "radiation fog" because the earth radiates heat into the clear atmosphere.

Fog forms in other ways, but the radiation type is most challenging to the night flier. It forms on such superb evenings that a pilot is tempted to fly locally without checking weather. If he returns to a field bound by fog he may have to detour fifty miles inland from the coastal area to find an alternate airport for landing. A clue to predicting ground fog is the trend in the temperature-dewpoint spread. If the gap is closing on a clear, cool night, ground fog is imminent.

A young sailor found that out as he took off from a Texas airport one night. The airport operator, watching from the ground, saw the plane heading toward a fog bank which rose to approximately one hundred feet AGL (Above Ground Level). The plane began a medium bank to the left and the pilot reduced power. A moment later, the aircraft struck the ground inside the airport boundary while in a 30-degree banked turn. The plane was a total loss, but the pilot and his passenger escaped with only serious head injuries. In the subsequent investigation these facts emerged:

1. There was no record of the pilot's having obtained a weather briefing before the flight;

2. The FBO had seen the fog bank rolling in and tried to warn the pilot on the unicom frequency. The plane was equipped for night and instrument flight, but the radio had not been turned on;

3. At the time of the accident a high, scattered ceiling covered the area, but visibility was estimated at one mile in fog;

4. The pilot, with a total of 125 hours, estimated his instrument time at 5 hours (under simulated conditions). In a written statement he said: "I cannot imagine what caused the accident because I had been trained and told that if I ever hit a fog bank along the coast on take-off, to climb through it, because it usually runs out at a couple of hundred feet of altitude";

5. The FAA accident report listed temperature and dewpoint at the time as 68 and 66.

Those five transgressions of rules and basic flying technique were committed in about the same number of minutes.

Unforecast weather is a hazard in cross-country flying at night. Lowering ceilings and clouds escape notice until bright red and green glows suddenly surround the wingtip lights. Cloud masses are nearly impossible to see on a dark night unless they reflect ambient light from cities below. Inside a cloud a flashing strobe or rotating beacon encircle

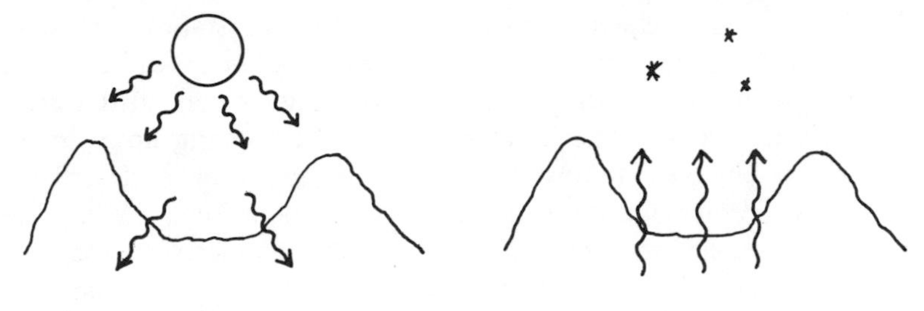

(A) Sun warms earth during the day.

(B) After sunset earth loses heat to clear sky by radiation.

(C) Moist air near ground is chilled below dewpoint, forming fog.

62. How ground fog forms on clear night.

the plane with dizzying closeness. Turn them off to prevent vertigo and execute a 180-degree turn back to clear air.

A landing light quickly reveals precipitation forming at your altitude. If outside lights are dimming, flick on the light and look ahead. Snow showers may barely cut down ground visibility but show up brightly in the glare of the light. Strokes of lightning as far as a hundred miles away silhouette cloud formations and warn of violent thunderstorm activity ahead.

The sight of distant lightning may also signal another disturbing phenomenon. It is a blue light radiating from propeller tips or engulfing the wings. The eerie glow is an electrical condition called "St. Elmo's Fire." This frightening effect was named during the fourth century when Spanish sailors saw the ship rigging glow with ghostly illumination and attributed it to the wrath of St. Elmo. It is generally considered harmless to aircraft, except for disruption to lower radio frequencies (especially ADF) and possible loss of VOR (VHF Omnirange) reception during severe attacks. Not only is it visible on the aircraft skin, but it may enter the plane and travel erratically inside the cabin!

St. Elmo's worst effect is usually precipitation static (or "corona discharge"). It is an accumulation of electrical charges on the plane as it flies through clouds, rain, dust, snow or fog, but it may also occur in clear air between clouds. The audible signs are a popping noise in loudspeakers or headphones followed by a sound that's been described as "frying grease." When the aircraft increases speed through the air, noise may build to a musical tone until it finally reaches an electrical roar which kills radio reception. In the early days of low-frequency electronics, pilots would slow down their aircraft during bouts of "P-static" to restore navigational signals. Today's VHF receivers are far less susceptible, but not completely immune.

United Air Lines considered it menacing enough in 1936 to send scientists aloft to discover its source. They learned that static electricity builds up thousands of volts on the aircraft skin which leak off in short pulses to the surrounding air. Those pulses emit a flutter of radio static. The cure is a series of small wicks which trail from various aircraft surfaces. Thousands of tiny points at the end of each wick allow static electricity to divide and silently leak away.

Weather at home

Since weather is a crucial item in night flight, you may want more than a routine briefing from a Flight Service Station (FSS) or weather bureau. Despite official urgency about securing a thorough picture of weather conditions, it is not always easy to obtain one on the phone. An excellent way to supplement the briefing is through specialized weather broadcasts picked up at home or office by simple radios.

An excellent accessory for a night flier is a radio which receives 24-hour weather broadcasts. Two services are available in much of the country, and each sends out a specialized version of current and forecast weather. Listen to both services and the total picture is remarkably detailed and complete. The equipment to tune these bands is widely available on conventional shortwave or multiband receivers. These receivers cannot supplant a weather briefing from official sources. You need vital, last-minute data from the FSS or WB. But a weather receiver at home helps form the decision about a night flight hours earlier in the day. The radio warns if the outlook is poor, before you make elaborate preparations or travel to the airport. Radio is especially convenient in areas where there is no direct line to a Flight Service Station or weather Teletype printer at the local airport. Here is a detailed look at the two most important weather radio facilities:

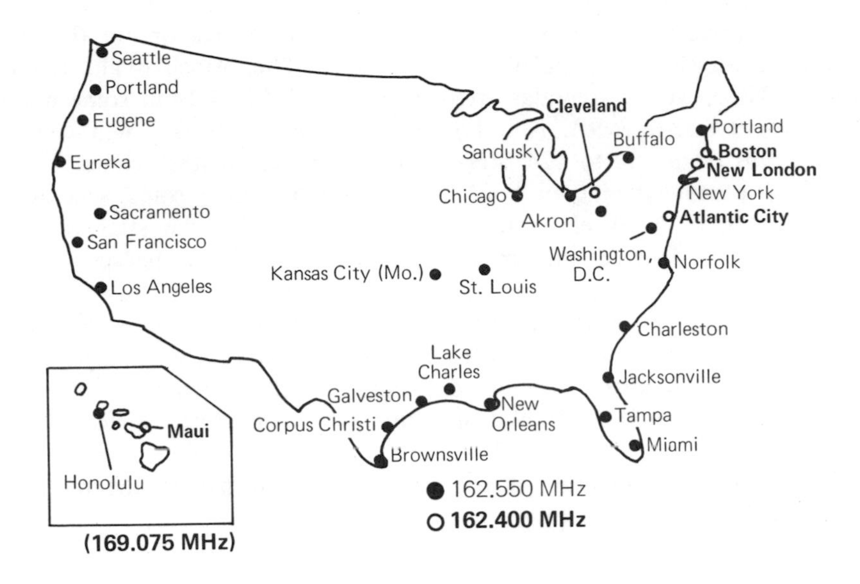

63. Network of VHF weather stations.

VHF-FM weather. This network, operated by the National Weather Service, is a group of stations which reach about 70 per cent of the American population. (VHF-FM means Very High Frequency-Frequency Modulation.) The stations broadcast an unremitting flow of weather reports around the clock, with the heaviest concentration along the coasts and the Great Lakes (see Fig. 63). When the service grew in the 1960s, it catered to the pleasure-boat owner, but its content is just as valid for pilots.

VHF-FM weather is far more detailed than the broadcasts of regular radio or TV. There is a complete regional and local weathercast on tape which runs about five minutes, then repeats. Every two or three hours the tape is refreshed with late information. A typical broadcast begins with an overall picture of the local area and adjoining states, with detailed observations of wind, visibility and sea conditions. If precipitation is seen on the radar screen up to about 150 miles away, its movement, size and direction are reported. The announcer continues with a marine forecast, the local picture, and a five-day outlook. The report is garnered from nearby weather bureaus and observations by the U. S. Coast Guard. That's the routine program. Since VHF-FM stations are manned by live personnel, an announcer can break in at any time with emergency information. A fast-moving squall, a heavy snow warning or other hazardous condition is quickly aired.

To receive these reports radios must be able to pick one of the two frequencies assigned to the VHF-FM service. The principal channel is 162.55 MHz, with a secondary frequency of 162.40 MHz in some areas (to prevent interference). A radio equipped for the VHF "high band," which runs from 150 to 174 MHz, can tune either weather channel. Instead of a conventional, or continuously tunable, dial, some sets have *crystal* control. A switch replaces the dial for tuning the station. If a crystal is used, its frequency must be compatible with the nearest VHF-FM channel.

The range of these weather stations is about 40 miles from the transmitting points shown in Figure 63 if you listen with a small, built-in whip antenna. Distance extends to about 75 miles when the station is high—like the one atop Mt. Wilson which serves Los Angeles. Ranges of 60 miles are common around New York City from a station atop Rockefeller Center in Manhattan. A rooftop antenna attached to the receiver significantly improves reception.

TWEB weather. Another service you may hear in home or office is TWEB, Transcribed Weather Broadcast. It operates continuously over Low- and Medium-Frequency stations which also serve as navigational aids. (These are the radio-beacon stations picked up by an ADF receiver in the 200–415 kHz band.) Since inexpensive portables for the home may include that band, the broadcasts can be heard by anyone. There are nearly a hundred stations equipped for TWEB in the continental U.S. as shown by the chart in Figure 64.

TWEB continuous broadcasts are strictly for aviation use. The announcer, for example, opens with, "This is Newark-Elmira Radio." Since he's identifying two stations, it means that the identical text is being transmitted from two locations. Note that these stations, Newark (EWR) and Elmira (ELM), are connected with a dotted line to indicate this linkage. Many such tie-ins are located throughout the country.

A TWEB broadcast describes weather within a 250-mile or greater radius of the transmitting station. It begins with a synopsis of general conditions and continues with flight precautions, forecasts, winds aloft, advisories and pilot reports. The announcer gives conditions along specific routes between cities and follows with hourly sequence reports from airports as far as four hundred miles away. The reports are especially valuable because they are current—observed within the last hour—and yield a dynamic pattern of weather movement. If you know weather-symbol shorthand, you can jot down sequence reports as they are spoken and get a grasp of weather trends in your region.

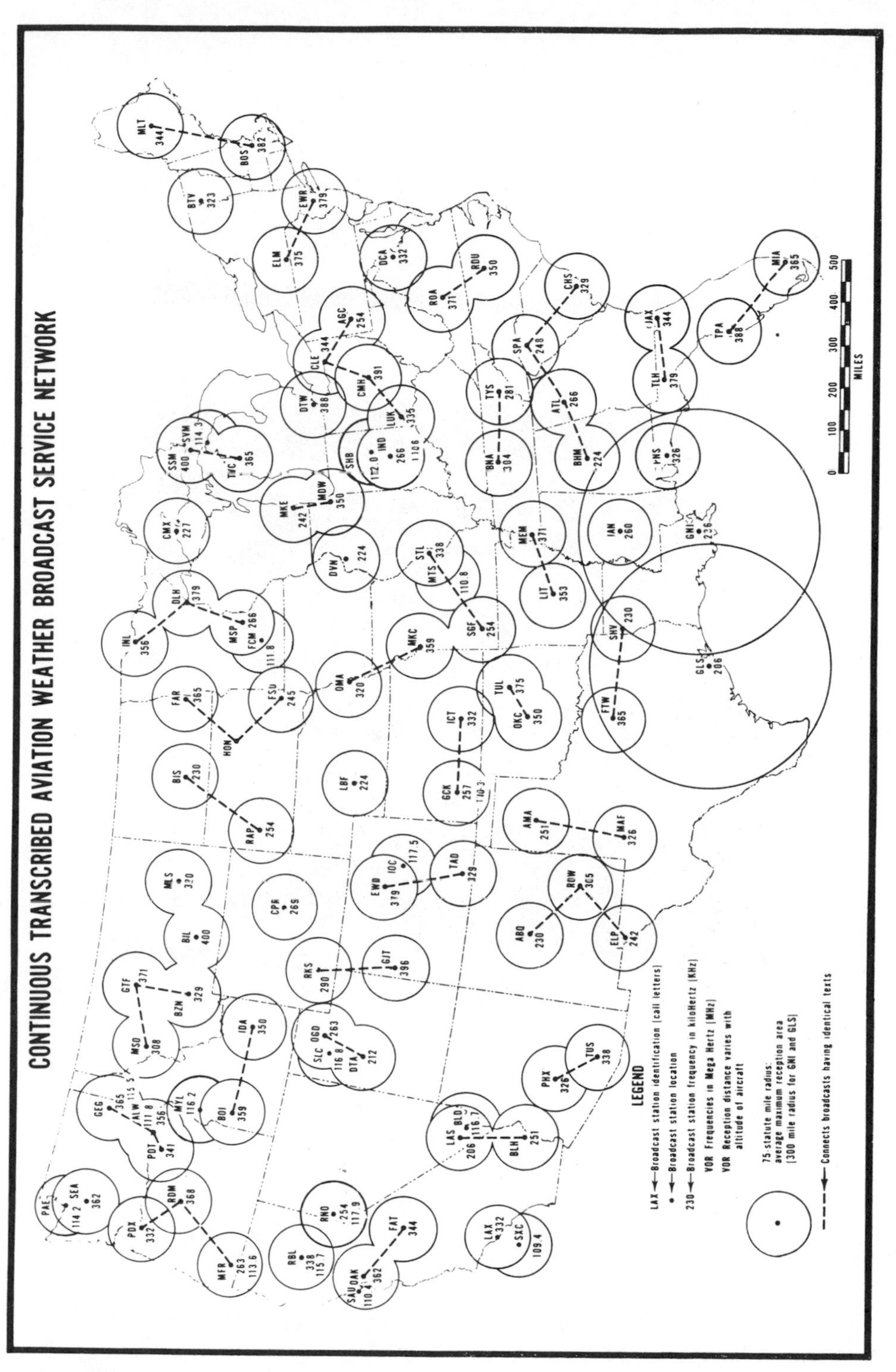

64. Stations, frequencies and range of transcribed weather broadcasts (TWEB).

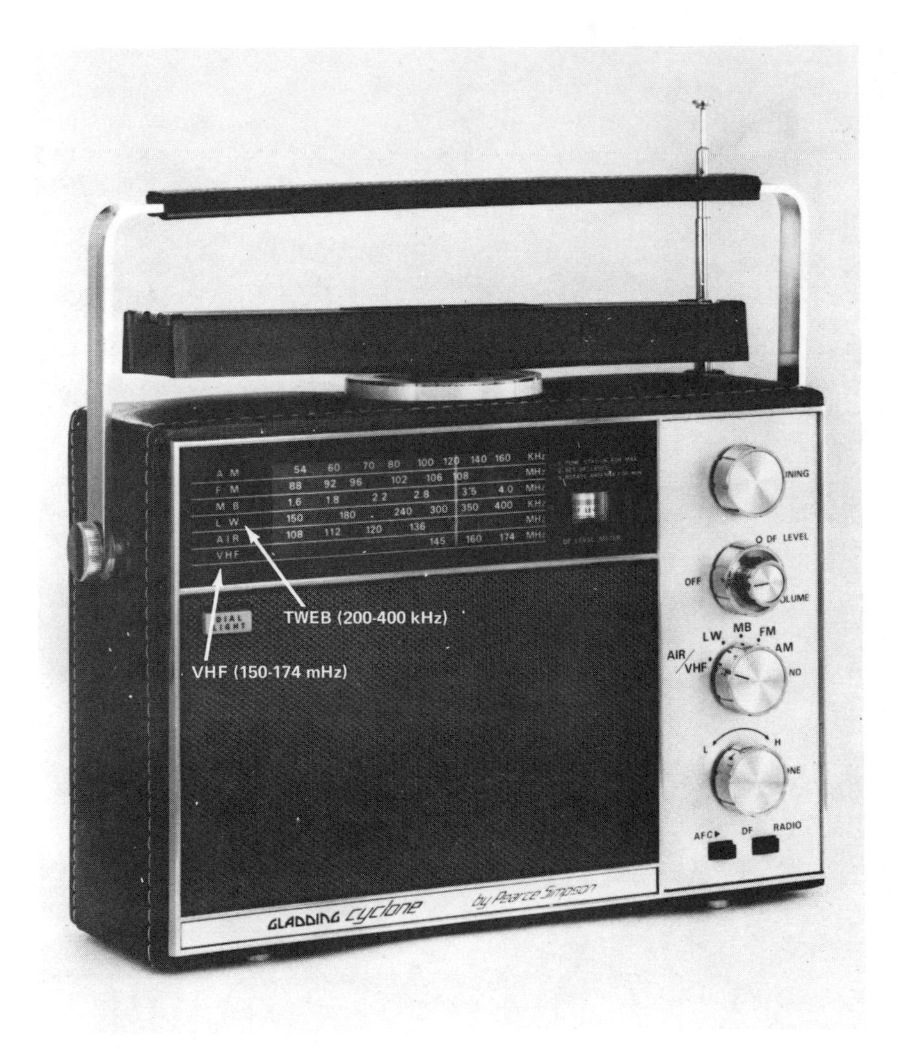

65. Single radio picks up both continuous weather services; TWEB and VHF-FM.

If you live within the circumference of any circle shown on the TWEB chart, the broadcast should be received on the built-in whip or loop antenna provided with the set. On the fringes or beyond, an outdoor wire between 30 and 100 feet long intercepts more signal. The radio, however, must have a provision for connecting an external antenna. The range of most stations on the chart is a 75-mile average maximum between transmitter and receiver (each circle is 150 miles across). Two powerful stations along the Gulf Coast, drawn with larger circles, radiate as far as 300 miles to cover the open sea.

It is possible to obtain single receivers that pick up both VHF-FM and TWEB services and provide a highly detailed view of current and forecast weather. A phone call to flight service or weather is still standard practice, but a weather radio at home brings an excellent preview of conditions aloft—without the risk of "going up to take a look."

First night flight

6

The first night flight recaptures the rare excitement of a first solo. But a widening rift cuts between nighttime and daytime fliers as the ground falls away. At the end of the runway the plane penetrates a wall of blackness where, accident reports suggest, a pilot with no night training encounters his first difficulty. In the classic case, the aircraft rolls safely down the runway and becomes airborne. With engine roaring and nose pitched up, the pilot senses he is climbing straight ahead. A dip of a wing goes unnoticed and the plane commences a gentle turn. The bank steepens until ground lights appear in a side window. The pilot instinctively hauls back on the wheel and triggers a stall at the worst time—nose high, airspeed low, close to the ground. Eyewitnesses are baffled by the simplicity of it all. The evening was clear, the plane took off, it made a turn and crashed.

A night checkout with a flight instructor lessens the danger. It's mandatory for pilots licensed after 1972 who want to fly at night, and recommended for the hundreds of thousands of pilots who elude the requirement by holding earlier licenses. A poor substitute for a checkout is shooting touch-and-goes at dusk and continuing them beyond sunset. Since the light wanes by imperceptible degrees, at some point the pilot mistakenly declares himself a night flier. Another risky introduction to night flight is a product of poor flight planning: when the sun sets before the plane reaches its destination.

How the hours of a night checkout are divided is left to the flight instructor. There is no syllabus of techniques or special maneuvers. Almost invariably the first hour is devoted to take-offs and landings. When the student is proficient in the traffic pattern, the instructor may venture beyond the airport and point out features of the local area. In the final phase of the checkout, a dual cross-country may be flown to tie together the skills of night flight. After the checkout, the student should feel sufficiently secure to solo around the pattern or fly to nearby airports. He is also current to carry passengers.

Instructors vary considerably in their approach and treatment of a night checkout. The material which follows, therefore, is not intended as a rigid course of instruction, but a synthesis of practices and judgments gleaned from professional pilots and instructors engaged in the daily task of preparing low-time pilots for night flight.

Preflight inspection

A pilot from Fort Yukon, Alaska, became an authority one evening on the value of nighttime preflight inspection. Soon after he took off on a local pleasure flight, he felt poor control response and decided to make a precautionary landing. During the approach the airplane swerved and collided with trees, but the pilot was not seriously injured. The cause of the accident was painfully apparent: He had taken off with a 25-pound concrete block tied under the wing.

Careful probing over the aircraft with a flashlight during a preflight inspection would have prevented the error. The light thrust into a tail cone or along the hinge of an aileron quickly confirms their condition. Play the light over large surfaces—fuselage, wings and empennage—to spot wrinkles, bends or damage that easily go unnoticed in the dark. Preflighting with a flashlight isn't much more difficult than performing the same chores in daylight. The narrow beam of light, in fact, may draw your attention to trouble.

Be doubly certain about items that could interfere with take-off. During the walk-around, grip each control surface—aileron, elevator and rudder (unless placarded otherwise)—and run it full travel. See if the yoke or stick responds in the correct direction.

Try every exterior light while you're outside the plane; position lights, landing light(s), anti-collision beacon or strobe. Don't turn all electrical switches on at once, because the heavy power drain might discharge the battery so the plane won't start, especially on a cold night.

67. Preflight with flashlight.

After turning on the master switch, operate each light switch on and off as quickly as possible. Never turn on the landing light for more than a fast flick of the switch if the engine is not turning. A helper expedites your lighting check by standing outside and calling off lights as you momentarily turn each switch on and off. After outside lights are verified, each switch or control for interior lighting can be tried after the engine is running to ease the battery drain.

Since so many night accidents are related to fuel exhaustion, this area is a highlight of a preflight inspection. Verify the amount of fuel by a direct look into the filler neck leading to each tank. If the rays of a flashlight don't show the fuel surface, stick a finger into the tank. At a fuel stop, watch the line man screwing on the gas cap. Some caps have a butterfly-shaped lock that can confuse a man who's new to the job.

68. Check overhead dome light; it can serve as a backup.

The missing gas cap is a notorious nighttime hazard. Gasoline disgorges through an open filler in flight at astounding speed. One professional pilot learned this one night soon after taking off from an airport in Texas. The airplane, a Cessna-206, had just been refueled in its right tank to bring the total fuel aboard to forty-five gallons (about half the plane's capacity). The pilot took off on the left tank and flew until it ran dry. No matter—he switched to the full right tank to continue the flight. Within moments the engine failed and the plane was substantially damaged in a forced landing. The pilot's last words on the accident report

69. Inspect fuel level in tank directly by eye.

were: "After landing I found the fuel cap off the right tank." His total distance from take-off to engine failure was eighteen miles. It is more difficult to check gas caps in high-wing planes, but fetching a ladder is worth the bother.

Missing oil caps also escape notice at night. The churning engine may splatter oil over the windshield in flight and lead to a treacherous landing. If oil escapes unnoticed, which easily happens in the dark, an engine seizure may be minutes away. One reason for the missing filler cap occurs when the oil is changed. The mechanic brings the oil level to nearly full, but is distracted by some other problem. He never returns to the plane and the oil cap dangles on its chain. If an accident occurs, the ultimate blame rests with the pilot, who failed to spot the missing cap during his preflight inspection.

When every item outside the plane is secure, check the ramp area for people, animals, ladders, trucks, ditches, construction apparatus and other aircraft. If none presents a threat to taxiing, it's nearly the moment to start the engine.

Taxiing

Don't touch the starter yet. FAA regulations specify that an airplane may not be operated during the period from sunset to sunrise unless it has lighted position lights. (In Alaska it's time when a prominent unlighted object cannot be seen from a distance of three statute miles or the sun is more than six degrees below the horizon.) The rules also say if you're "in dangerous proximity" to night-flight operations of an airport, the aircraft must be clearly illuminated and its position lights on, or it must be located in an area marked by obstruction lights.

To comply with the rule, turn on the position lights, but, to save the battery, wait until the engine is running before switching on other lights. Yell a hearty "Clear" through an open door or window before turning over the engine, because people or animals are often unaware you're about to spin the prop. After the engine is turning, switch on the anticollision light and instrument-panel lighting, then adjust intensities to the lowest practical level. It is a courtesy to pilots operating near you to leave off a high-intensity strobe during ground operations. The white stabs of light reduce night vision at close range. (Remember to turn it on later.) If the plane has a single landing light, turn it on to illuminate the taxi to the run-up area.

As the aircraft responds to the throttle, look ahead to find the line of blue lights that edge the taxiway. Listen on the radio for the traffic flow in the area. If you're on an uncontrolled field, other aircraft will probably broadcast position reports in the traffic pattern on the unicom frequency. In controlled airports, you'll be under instructions from ground control. (Forget to turn on your lights while rolling and you may get a reminder from the tower: "I don't have your navigation lights, sir.")

One flight instructor sums up the next step this way: "I must continuously caution my students against taxiing too fast at night. They all do it!" He suspects that a pilot new to night flying operates with a greater sense of urgency and speed. With the loss of normal visual cues the beginner finds it more difficult to judge the airplane's speed over the ground in the dark. Instead of terrain moving past the eye, the references are points of light which may confuse, rather than aid, the pilot's perception. Taxi speed should be no greater than the plane's ability to stop within the distance disclosed by the landing lights.

Many small airports have lighted runways but no taxiway lights. Move slowly here because it's easy to run into soft ground or a ditch on a dark night. Some airports have lighted taxiways with a nocturnal surprise: As you roll along the taxiway is suddenly plunged into darkness. To conserve power late at night, an automatic timer cuts off the taxiway lights (although runway lights may burn brightly). If this should happen, reduce speed and follow the landing light.

It's easiest to taxi at a field with an operating tower, because ground control announces wind direction and assigns the active runway. If the field is strange and no other airplanes are about, you'll probably need directions through the bewildering maze of ground lights. The controller can guide your plane through each turn to the runway, but you must say something like "Instructions, please," during the call-up. Some controllers become testy if you don't track the yellow centerline of the taxiway. It disturbs their orderly traffic flow—and you may clip an invisible aircraft parked in the dark.

Taxi most cautiously at an uncontrolled field because of the uncertainty about the active runway. Unicorm is probably retired for the night, so you must make the decision about which way to take off. Part of the choice is already clear if there is only one lighted runway. (It's hazardous to attempt take-off from a dark runway.) However, don't trust a tetrahedron or wind tee as an absolute indicator of take-off direction on the lighted runway. Some airport managers tie down the indicator after dark in the direction which favors the prevailing wind during good weather.

The windsock is an excellent reference. The only problem is seeing it from the run-up area. Chances are it's lighted, but its direction may be deceptive as you sit in the aircraft near ground level. Since wind is so crucial to take-off length and climb-out angle, it is worth a delay and detour to taxi toward the sock until you are certain of actual wind direction. If the sock is blowing at right angles to the active runway, favoring neither direction, listen for other aircraft in the area to determine the direction they are using. It helps eliminate the harrowing possibility of two aircraft at night approaching the same runway from *opposite* ends. Scan the traffic pattern overhead for other aircraft as you taxi.

Run-up

A nighttime run-up probably takes twice as long as during the day. If a pilot commits an error that ultimately leads to an emergency on take-off, the chances are good that he could have caught it during the run-up. The first defense is a printed checklist—not a contrived acronym like CIGAR, but a manufacturer's detailed checklist. Controls that are highly visible during the day—trim tab, carburetor heat, wing flaps, fuel selector (or an open door)—easily escape notice in the dark. Any one item could cause a take-off accident, especially at night.

As you run down the checklist, poke the flashlight around the cockpit to confirm the settings of knobs, needles or levers hidden from the aircraft's cabin lights. (Red light is preferable to help your night vision.) If the checklist is not fastened to the plane, hold it (or the owner's manual) under the cabin light to read each entry. When you're certain every item is ready for take-off, reduce all lights to the lowest possible level.

The next step is optional. It is a full-throttle, static run-up just before rolling into take-off position on the runway. The magneto check during the regular run-up may prove nothing's amiss, but in the next few moments plane and pilot are thrust into the vulnerable condition of a limited-visibility climb-out into darkness. It is no time for less than full power. The engine should take the smooth application of power on the ground without hesitation and reach nearly full rpm. The static run-up is also insurance against carburetor ice on a damp, cool night. A touch of carbuetor heat at high rpm should clear any ice that formed between run-up and take-off.

Take-off

Roll onto the centerline of the active runway. Nothing less than exact center is accepted by many instructors, because this position could prevent an accident further down the runway if the plane drifts in a crosswind. The runway lights quickly disappear after lift-off and leave you little reference that the plane is drifting to the side.

Don't apply the throttle until two items are confirmed: that you are not lined up on the wrong runway, or about to take off from a taxiway. If you taxi to the wrong runway at a controlled field, don't depend on the tower operator to shout a warning on the radio. He may be a mile away, unaware of your mistake. A glance at the magnetic compass or directional gyro helps tell if you're on the correct runway.

Taking off on a taxiway is another hazard. At airports with acres of hard surface, it is difficult to distinguish the runway from the taxiway. The revealing cues lie in colors and lights. Taxiways are rimmed with blue, the runway is bordered by white lights. Check for green lights astride the runway threshold, or the red lights of a displaced threshold. The runway centerline ahead should be a broad, white broken stripe, not the thin, yellow line of a taxiway.

Keeping the airplane straight down the runway on take-off is usually easy. The silhouette of a distant hill or a lighted object provides a reference ahead. Also, the landing light illuminates the runway centerline. On the darkest night, the strings of runway lights on either side supply enough reference while accelerating down the runway. Some students have trouble with the plane's tendency to veer left as full power is applied. The turn is followed by an overcorrection to the right and the plane S-turns down the runway. One instructor has a beginning student apply partial throttle (about 1,500 rpm) to get the plane rolling fast enough for good control response in the airstream. After the student can hold the airplane straight he applies full power.

Students often roll too long and too fast on a night take-off because they believe there's safety in speed. High velocity may reduce chances of a stall, but it also invokes penalties. Too much runway is consumed because lift-off happens too late. And the extra time on the ground could have been spent climbing to clear obstacles. The airplane should be rotated at the speed recommended by the manufacturer. Don't stare at the airspeed indicator at this critical time, but take quick glances and apply back pressure to place the nose in a take-off attitude. The plane should fly when it's ready.

After the plane breaks ground, verify climb speed on the airspeed indicator. This is important on a night take-off. As the ground falls away, there is little reassurance that the plane is really ascending. It is not unusual for a beginning student to rise from the runway, lower the nose and accelerate to cruise speed while the plane is at dangerously low altitude. Besides monitoring airspeed, scan the artificial horizon, altimeter and VSI (Vertical Speed Indicator) to confirm the climb. Look out at lights below to be sure the ground is receding. The ascent following take-off is a critical moment in night flying because there is little feeling of vertical motion.

What is the best speed after lift-off? Some instructors prefer the manufacturer's recommendation for maximum *rate* of climb. If the field is short and you believe the speed for maximum *angle* of climb is necessary to clear an obstacle, then reconsider the whole flight. A slow-speed ascent in the dark may be too hazardous to a pilot of average skill.

When the plane is established on climb-out, keep checking flight attitude both visually (outside) and against instrument readings. A night climb-out is a time of poor outside visibility, and it's risky to allow a wing to fall off without seeing it happen on the artificial horizon or turn needle. Maintaining a straight track (Fig. 70) off the end of the runway also avoids obstacles to the left and right of the flight path.

Opinions vary on when to turn out the landing light. Some professionals never even use it on take-off—to preserve their night vision. To a low-time pilot, the landing light creates two illusions on take-off. As the nose rises for climb, the lamp projects a powerful shaft of light ahead of the plane. It is as enticing to follow as the white stripe running down a dark runway. However, this could prove disastrous in the air. Since the landing light is fixed to the plane, it always shines straight ahead, regardless of the airplane's pitch. *Never* use the landing light as a reference for climbing out.

The second illusion with a landing light is that traces of haze in the air fool the novice into thinking he has flown into solid fog or low clouds. It can be frightening, because the brilliant light produces powerful reflec-

70. Maintaining runway heading to avoid hills and obstacles.

tions back to the cockpit from the slightest haze. Glance out a side window toward the ground or horizon and lights or other features below should confirm the actual clarity of the air. A similar effect happens on winter evenings when there's a scattered or broken cloud layer overhead. A brief snow flurry trapped in the landing light resembles a driving blizzard. Switch the light off and visibility returns.

A good time to turn off the landing light is shortly after take-off, as soon as the plane is climbing. The purpose of the light is to illuminate the ground, not the sky. Some pilots keep it on while flying all legs of the traffic pattern as a warning device (in addition to the anti-collision light). This is a matter of personal preference and pocketbook—the landing light of a small plane may have a total life of less than a dozen hours.

Your instructor should recommend the altitude for entry onto the crosswind leg. During that first turn banking wings blot out much of the outside world, so make a quick crosscheck of the instrument panel. Flying on the crosswind leg, glance down, out the side window, to find the two rows of runway lights. They serve as the primary reference for flying the pattern.

Landing

The downwind leg is more crucial to a good landing at night than during the day. The terrain is often a confusion of lights and its visual appearance difficult to interpret. The one certain reference is the runway lights. You know, absolutely, their height and your aircraft's altitude in relation to them. By disregarding superfluous lights around the airport and concentrating on correct altitudes, a descent is gauged against a known reference. Don't become distracted on downwind by turning on the landing light. Unless a tower calls for a light to identify your plane, it serves little purpose on downwind. As the plane reaches a point opposite the intended touchdown, reduce power for a normal descent (usually around 500 fpm).

A common nighttime accident results from flying into the ground short of the runway on a perfectly clear night after a straight-in approach. It takes its toll among low-time pilots and airline captains alike. The reason is that altitude is extremely difficult to judge over miles of blackness preceding the runway unless VASI lights give vertical guidance. A pilot can almost completely avoid this hazard by always flying a conventional traffic pattern with a downwind leg. It compels you to fly parallel to the field and receive the reliable reference of runway lights. The straight-in approach is a tempting timesaver, but one that multiplies the chances of an approach accident.

After the plane turns to final approach, descending at a steady rate, turn on the landing light. Use normal approach speed. A common complaint among instructors is that students fly faster than necessary on final approach, coupled to a tendency to fly too high. Speed and altitude yield a secure feeling, but also raise the danger of a long landing and overshoot. One instructor doesn't castigate his students for adding five mph to a night approach, but he insists they don't add five more for *every* person aboard.

A slightly steeper-than-normal approach angle is favored by many professionals to be sure of clearing obstacles. They also use some engine power throughout a night approach, then chop it completely on touchdown, holding the landing roll to a minimum. Partial throttle also helps a pilot compensate more quickly for errors in judgment during the approach caused by reduced visibility or unfamiliar lighting.

The runway edges may sometimes appear higher than the adjacent terrain. The illusion is created by runway lights, which apparently lift the field and tempt you to flare out too soon. This is avoided by not staring at runway lights, but scanning the general area. If you suddenly see one or more runway lights disappear, *immediately consider* aborting the landing, applying full power and climbing for a go-around. You may have slipped too low and a tree or other object is blotting out several runway lights. After commencing a good descent from the proper altitude toward a public airport, i.e. one with clear approaches, you should encounter no surprises along the final approach path. If there are obstructions (as there are around many airports) red beacons should make them clearly evident.

When there is no vertical guidance to a runway (like the VASI light) a glide path can be estimated by the apparent movement of the runway in the windshield. If it moves higher in the windshield, you're too low and may undershoot; if the runway moves lower, you're probably too high. One pilot cannily recommends his own brand of VASI: If you look at the third and fourth runway lights beyond the threshold, and keep them the same distance apart, you'll maintain a steady glide path and touch down near the far light; if the field is short, use the second and third runway lights.

A few moments before the plane comes over the threshold, the landing light floods the runway surface. Don't attempt to put the airplane into that widening pool of light ahead of the nose. As the plane glides to the runway, two rising rows of runway lights flash by and outline the location of the runway surface. This is the primary visual reference for breaking your glide and starting a flare. Judging by the landing light alone distorts the perspective. As the nose rises during the flare, the pattern of light continuously elongates and changes position ahead of the airplane. During any normal landing, therefore, crosscheck between runway and landing lights for the composite view you need for touchdown and roll-out.

Do not land at night if you cannot touch down in the first third of the runway. The accident risk rises rapidly if a go-around is initiated near the ground in the dark. To apply full power, retract flaps and start the airplane climbing may prove too demanding at low altitude. A decision for a go-around should be taken early on a night approach.

As the plane slows to taxi speed after your first night landing, you'll breathe a sigh of relief (and feel a touch of pride). But don't relax yet —finding the taxiway is a challenge. Turn-offs are sometimes marked with lighted signs and rimmed by blue lights, but seeing them is another matter. Avoid embarrassment by watching the runway centerline ahead for a yellow stripe that arcs away toward the taxiway.

Simulated emergencies

In the beginning hours of night flying, an instructor may stage a number of emergencies. The most popular ploy is to fail the landing light. Depending on an instructor's ingenuity, it happens in several ways. He may say on downwind, "Don't use the landing light this time." It's disconcerting to a student because the shaft of light becomes a visual crutch that seems essential to the last minutes of flight. But the feeling dissipates after several no-light landings. Runway lights give the essential guidance on either take-off or landing. Also, wingtip lights cast a slight reflection on some runway surfaces just before it's time to flare. More adventurous is an instructor who pulls electrical circuits with no warning. He may do it on take-off to see if a student can safely continue the flight.

In a thorough checkout the student should demonstrate an ability to fly solely with a flashlight. One instructor initiates a cockpit drama by killing the lighting system a few miles from an airport, then expects the student to land. As the lights go out, the student gropes for his light and unwittingly lets the plane drift toward an unusual attitude. The instructor points out the first error: A flashlight must be instantly available —clipped to a pocket, suspended by a lanyard from the neck, or held on the lap.

The second crisis soon emerges. During the approach the student plays the light on the instruments to see airspeed and altitude. If the night is dark and he's in a turn with no outside references, he swings the light to check the artificial horizon. So far, so good.

"Emergency go-around," shouts the instructor. The student suddenly has his hands full of flaps, throttle, carburetor heat and flight controls. He hasn't enough hands, so he lays the flashlight on his lap. As the airplane staggers, the flashlight points to the rudder pedals. He tries to poke the flashlight between his knees, but now the beam shines on the cabin ceiling.

The instructor rescues the student by handing him an L-head flashlight. Placing the lower end between his thighs (Fig. 71), he demonstrates how it shines directly on the instrument panel. Then he grasps the light and swivels it left and right in sweeping motion. It can be aimed anywhere on the panel and remain fixed in position with no hands. This instructor believes the ability to fly by flashlight is as important as any other item he teaches on a night checkout.

The L-head flashlight isn't the only solution. Work out your own hardware if you wish, but try it in the cockpit under actual night conditions on the ground. Some pilots prefer to carry a penlite flashlight in a shirt pocket for instant use, and a regular-size type in a flight case. The larger light is less accessible, but the pocket flash is ready for any emergency.

71. L-head flashlight permits hands-off operation.

After take-offs and landings, you will probably leave the traffic pattern. Instructors vary in what they teach next. Some point out familiar daytime landmarks and explain how they appear at night. Others challenge you to approach an airport located near a body of water or unlighted countryside to see if you maintain control while maneuvering to land. One instructor regularly takes night students to a large airport in a heavily populated suburb. Landing at this field, which is surrounded by thousands of moving and fixed lights, is made even more intriguing by a right-hand traffic pattern. An instructor may demonstrate how to check for the first traces of ice by playing the flashlight beam along the leading edge of the wing. If your instructor has you land at an airport with parallel runways, watch out for a common error: lining up with the wrong one.

Some instructors give no further airwork. After circuiting the pattern and flying the local area the pilot is considered checked out. He has accomplished the correct number of landings and hours to meet official requirements. Such a checkout is legal, but incomplete. If a student cannot demonstrate an ability to hold the plane in a steep turn *solely by reference to instruments,* his night-flying ability is seriously in question. One West Coast flight school demands that students perform several 720-degree turns at a 60-degree bank for night qualification. To be sure the student can maintain control, the plane is flown over water, several miles offshore. During the turns, city lights afford a good horizon through part of the maneuver, but half the circle must be done in darkness. To prevent disorientation during this confusing situation, it is usually safer to fly the maneuver with reference to instruments.

Beyond the pattern

A beginning night pilot may fly his first true cross-country flight alone. Take-offs, landings and local flying are usually enough to satisfy flight instructors that a pilot can handle the airplane. But before traveling beyond the airport area, consider these suggestions to keep the early hours of night as safe as possible:

Fly the first solo hours in excellent weather. A clear sky and bright moon remove the startling impact of losing outside references in the darkness. Don't complicate take-off and landing by flying in anything more than a gentle crosswind. If the wind picks up while you're aloft you'll see the plane's drift while in the approach to landing. Runway lights clearly slide sidewise unless you're correcting for drift.

FAA adds other cautions about night flying. The student should not make his first solo when there is a great deal of other traffic. Although nighttime collisions are extremely rare and other planes easy to see, be alert in the pattern. Never perform aerobatics at night. The agency also recommends that a student remember the terrain adjacent to the airport. He might be able to pick out a suitable field in the event of an emergency.

Memorize the position of all controls. Practice on the ground with eyes closed until you can locate every switch, knob and lever. If this is not practical because you fly different aircraft, call out each switch and operate it during the ground run-up as a refresher.

Know any idiosyncrasies of your airplane that could prove surprising in flight. Some light planes shudder shortly after take-off as the airframe resonates or "oil cans" through certain rpm regions. It sounds ominous in the dark if you don't expect it.

After several hours alone in the night sky and short hops to nearby airports, your ability to fly the plane under varying light conditions is probably improving. The tranquility of the night compels you to think about cities and airports hours beyond the horizon.

You may be ready for the first long cross-country after dark.

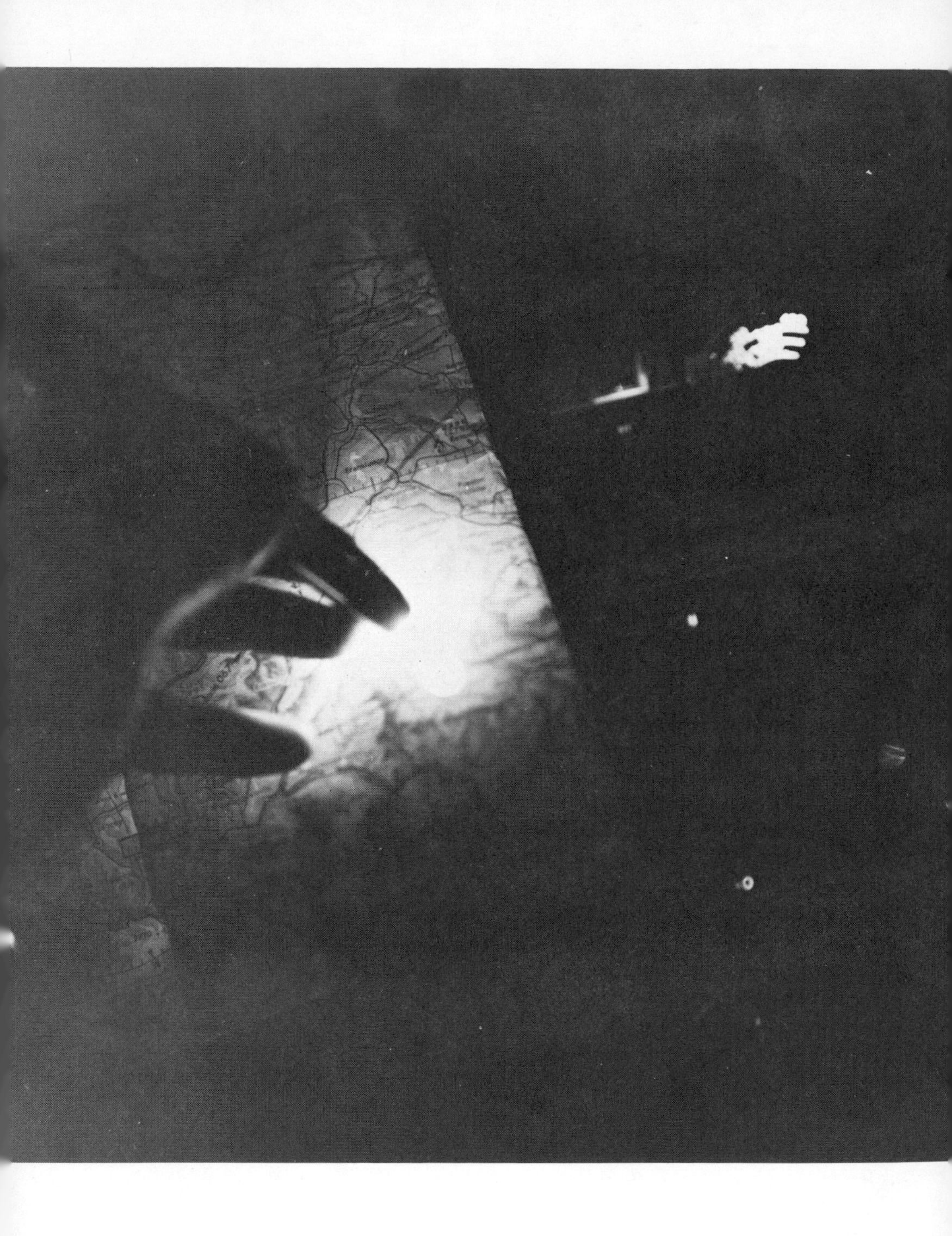

Nighttime navigation

7

When Alcock and Brown made their record-breaking nonstop flight across the Atlantic in 1919, nighttime navigation had more than its share of uncertainties. Brown brought along a ship's sextant to shoot course information from moon and stars. As he was about to take sightings that historic evening, the Vickers-Vimy biplane flew into a thick fog. Over the roar of Rolls-Royce engines, Brown shouted to Alcock to go higher. It wasn't until they reached 12,000 feet that the stars finally reappeared to guide them.

Today the challenge of getting from Point A to Point B in the dark is vastly simpler, for the tools of navigation are greatly improved. The three navigational techniques of day flying—dead reckoning, pilotage and radionavigation—are available for the night flier, but each system is somehow modified after dark. Pilotage points are severely limited, shifting winds distort dead reckoning, or stray radio signals lure a plane over false courses. Because there is no single, foolproof method for tracking accurately over the darkened ground, all three systems should operate in concert. Crosschecking between them prevents a gross navigational mistake.

Whopping errors can strike the unwary pilot. One student recalls how it happened to him on his first night cross-country with an instructor. After taking off into the setting sun from a Boston suburb, the student planned to follow a southwesterly course toward New York City. He dialed an omni radial leading to the first VOR on a carefully plotted course and made his course adjustments. Then minutes after take-off, as the sun's disk disappeared behind the horizon off his wingtip, the student tried to center a recalcitrant VOR needle. When the towers of Boston grew larger in the waning light, the instructor's glare affirmed an obvious conclusion: The radio was defective and the VOR course was leading the plane in the opposite direction. The student tried to plead innocence by saying the radio malfunction wasn't his fault. The instructor roared back, "Since when does the sun set in the east!" The lesson was painfully clear: Crosscheck your navigation against any available information, sun and stars included!

Compass and pilotage

In writing about his 1927 flight, Lindbergh said, "It is almost as essential for the liquid compass to work as the engine to keep running." The remark still carries wisdom. As the most basic device for dead reckoning, the compass is a reliable, uncomplicated instrument. But the night wind is deceptive, and course corrections to compensate for wind drift are not easily ascertained. Telltale visual reminders—ripples on water, the tilting of smoke—are missing in the dark. The day's turbulent bumps are probably gone. Aloft on a clear night a pilot may be surprised at the huge rudder correction needed to keep a VOR needle centered in smooth air.

Another effect belies the wind's intensity. As a pilot holds a wind correction during the day, the crabbing aircraft points its nose to the left or right of the unwinding terrain below. The cue is more subtle at night unless you concentrate on the relative movement of ground lights ahead. Over dark terrain there is no apparent motion. If a correction isn't made quickly the plane may deflect far off course in remarkably few minutes. The cure is a winds-aloft forecast and a pilot's careful refining of his drift correction against known checkpoints along the way.

Since rivers, lakes and cities often stand out clearly at great distances, important checkpoints are visible to make pilotage possible at night. One of the best indicators in the night sky is an airport beacon. It takes some practice to make a positive identification. First, scan the horizon for a white or green glint, then slowly count five or six seconds until the other color appears. Confirm it with a few more sweeps so you're not misled by marine or other lights.

Some pilots beacon-hop on a cross-country flight. They select legs of the trip based on lighted airports as checkpoints, even though it lengthens the trip. Beacons are visible on clear nights at distances greater than most daytime checkpoints. One state, Massachusetts, boasts that a pilot flying over its territory "will never be more than seventeen miles from a lighted runway."

Flying beacon-to-beacon also improves prospects for an emergency landing. With an airport nearly always in gliding distance, some pilots feel that night flying is nearly the same as day VFR. Others argue that it still takes considerable skill to dead-stick into any field at night. To estimate your aircraft's performance, check its table of gliding distance vs. altitude. From a height of 10,000 feet, for example, a Cessna-172 glides about seventeen miles with no wind, which means it could be within reach of a lighted field much of the time in many parts of the country.

Highways are also popular for eyeball navigation. Like airport beacons, they lengthen the trip but are reassuring over sparsely populated territory in areas like Maine or the Pacific Northwest. Watch out, however, for the intricacies of the interstate highway system. Back in 1923, on a record nonstop flight from New York to San Diego, Macready and Kelley ran into low visibility outside of Dayton and found their way to Indianapolis by automobile lights on the highway below. They might not have done it today. Complex routes around most major cities are a challenge to pick out from the air. It's too easy to fly toward a town over a known highway, lose the road in the city glare, then fly out the wrong route. If you attempt highway flying, confirm the road with a compass heading and sectional chart.

Practical night flying beyond the local airport depends heavily on radionavigation. The continental United States is saturated with VOR and NDB (Non-Directional Beacon) stations to make navigation nearly effortless. Consult AIM before take-off to learn about off-the-air stations, and don't depend on a VOR whose audio identifier is missing (a warning that technicians are at work). VOR stations suffer occasional anomalies which waver the needle or cause signals to travel hundreds of miles beyond normal line-of-sight. These ailments, however, are not peculiar to night operation. The same cannot be said for low- and medium-frequency signals picked up by the ADF receiver.

"Night effect" and ADF

Soon after the United States installed its first radio range, in 1927, between New York and Pennsylvania, Haraden Pratt, an engineer, described a curious phenomenon. Aircraft could navigate accurately along the new electronic airways during the day, but something affected signals after dark. As a pilot tracked over the ground in a straight line, his navigational receiver indicated that the aircraft slid slowly back and forth across the course. The undulations grew worse with the setting sun and disrupted navigation for hours at a time.

"Night effect," as it's called, had been noticed in simple radio direction-finders as early as 1916. Pratt's study in 1928 cleared up the mystery. During the day, low-frequency radio signals hug the surface of the earth and travel via a *ground wave*. How far they reach depends on the amount of power at the transmitting station. Airport compass locators, for example, extend about fifteen miles, while higher-power stations in the low- or medium-frequency bands cover seventy-five miles or more. That's a daily, dependable figure.

The ground wave isn't the only signal emitted by a beacon station. The transmitting antenna compresses the signal close to the earth's surface, but part of the radio wave escapes toward the sky. It may soar hundreds of miles toward space until it strikes the ionosphere, an electrical mirror created by the sun's radiation. As the radio wave intercepts the ionosphere, it reflects downward like an electronic billiard ball (Fig. 73). That signal cannot return to earth during the day because the reflecting angle is too shallow. The signal spills beyond the horizon and escapes into space.

With the waning of the sun's rays, the ionosphere weakens and withdraws to higher altitudes. Now the signal must reach a higher level before it bounces and changes direction. As shown in Figure 73, the net effect is a steeper angle of reflection and the signal returns to earth. Night-flying aircraft, therefore, can receive *two* signals from the same station: the direct, or *ground* wave, and the reflected, or *sky* wave. This is what confounded early direction-finders. The advancing ground wave maintains a constant angle with respect to the earth's surface, but a sky wave strikes the turbulent underside of the ionosphere and tilts unpredictably before it speeds back to earth. In an aircraft receiver, the twin signals merge and cause "polarization error." The ADF needle swings erratically with the varying state of the ionosphere.

The problem is eliminated in newer navigational systems based on VHF signals. Signals in this band penetrate the ionosphere at shallow angles and almost never reflect earthward. For this reason a VOR receiver suffers no night effect. ADF, however, is still prone to the errors explained in 1928 by Pratt. Since it is so valuable to a night flier, consider how to cope with its anomalies. The ADF receiver is an excellent backup if the VOR fails; in some regions it might be the sole electronic facility.

The first clue to night effect is when it happens. Greatest impact tends to be just after sunset and immediately before dawn. The ADF needle wavers back and forth in an excursion linked to range; the fluctuations

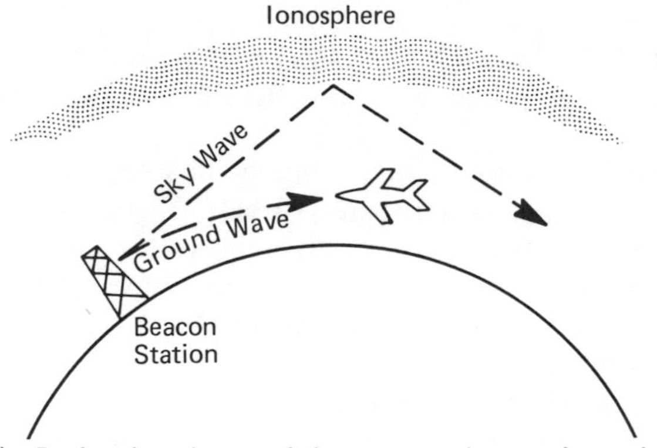

(A) During day, skywave is lost to space because ionosphere is at low altitude. Reception is via ground wave.

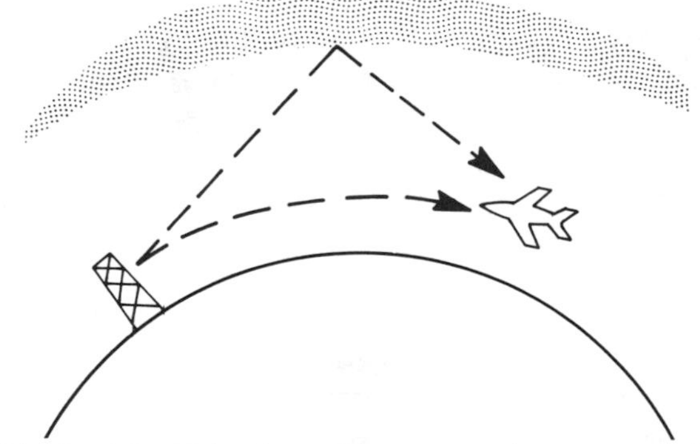
(B) At night, higher ionosphere returns skywave to earth, which may confuse ADF signal on ground wave.

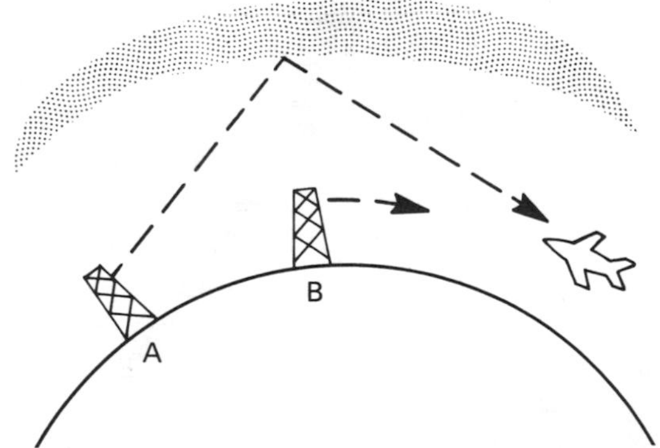

(C) Error is also caused by signals arriving over great distance at night. If station A and B are on same frequency, aircraft may incorrectly home on A, especially if B is out of range.

73. "Night effect" on ADF.

tend to deepen with distance from the transmitting station. If you must use that station, attempt to fly a bearing which averages out the needle swings.

When there is a choice of ADF stations, pick the one with the lowest frequency, e.g. 257 kHz rather than 379 kHz. Night effect diminishes as frequency is reduced. Toward the bottom end of the low-frequency band, signals increasingly hug the earth and generate weaker sky waves. Choosing a stronger station also helps to cut down night effect.

Always identify an ADF signal at night by listening to its audio in the speaker or headphone before switching to the navigational mode. Sky-wave signals have amazing vitality because they have little direct contact with the earth. It's possible to tune to the desired frequency but pick up a false signal originating hundreds of miles away. The precaution also applies to digitally tuned ADF. Such dials tune with great accuracy, but the receiver may still capture a distant, spurious signal.

Homing on commercial broadcast stations at night is trickier. Their powerful signals are attractive for homing, but station identification is too infrequent. As on ADF bands, sky waves bring in many signals which share the same frequency and cause bearing error. One clue is a whistling sound created as the two radio signals (near and far) mix and generate an audible tone. If you hear whistling, or more than one program simultaneously, don't rely on this dial setting for direction-finding. Night effect often weakens below 1,000 kHz, so try to select stations of lower frequency if you must home in on a standard broadcast station.

There is one more vagary of nighttime ADF. It happened this way to one pilot on his first night flight toward his home airport. A local broadcast station, conveniently located near the end of a runway, had always been an excellent homer during the day. Its powerful signal could lead the pilot within sight of the field from more than forty miles away. But on this evening the pilot sensed he was overdue, even as the ADF needle steadily indicated the general heading of home. He switched on the audio and, after hearing the station identification, discovered his "homer" was located several states beyond his airport. The hometown station was a "daytimer" that went off the air at sunset. When this happened, it allowed a distant broadcaster on the identical frequency to arrive from hundreds of miles away via the sky wave—and feed the receiver with false information. ADF is more critical to use at night, but the pitfalls can be handled by an alert pilot.

Compass locators

Major airports offer the ADF-equipped pilot an excellent homing aid to guide him over bewildering city lights to the traffic pattern. It's the compass locator associated with an ILS. A low-powered station, it emits a signal in the 200–415 kHz band which can be received on an ADF at a range of about twenty miles from the airport. It primarily furnishes an approximate five-mile radio fix for aircraft maneuvering onto the final approach path during instrument operations, but it helps VFR aircraft as well. Note that a compass-locator signal can be picked up while approaching from any direction and captures the ADF needle in conventional fashion. Be sure to allow for the fact that a compass locator is about five miles off a runway threshold.

§ **COLUMBIA REGIONAL (COU)** ***IFR*** 10SE **FSS:** COLUMBIA on Fld
889 H65/2–20(1) (S–92, D–125, DT–215) BL6,7A,13 S5 F12,18,30
U–2
Remarks: UNICOM 123.05 avbl.
ILS 110.7 I–COU Apch Brg 015° **LOM:** 407/CO
Columbia (L) BVOR 111.2/CBI/122.1R
VHF/DF Ctc FSS

74. Compass locator details given in AIM.

To flight-plan with a compass locator, look up the field in Part 3 of AIM before take-off. Assume it is Missouri's Columbia Regional, as shown in Figure 74. Find the letters "LOM," which represent Locator Outer Marker. This is followed by "407," the frequency to which you tune the ADF receiver, and "CO," the Morse code identifier. To determine the actual position of the compass locator, note the approach bearing ("Apch Brg"). In this example it is 015 degrees; if you approach the airport on that heading, you should arrive at the compass locator (and see a needle reversal) about five miles before the threshold of the instrument runway.

Don't fly near the airport on that approach bearing at night unless you clear this intention with the tower or the radar facility controlling the area. Even during good weather, heavy aircraft are vectored onto that course (the localizer) for their final approach. The nose-high, rapid descent of a big aircraft makes the route inhospitable to unknown light planes. Recall, however, that a compass locator is usable from any direction, so you may elect to approach on a different heading.

DF steer

On a sunny weekend it's not unusual to hear a lost pilot receiving homing instructions by radio. He's asked to call out landmarks, adjust his VOR receiver and describe the terrain. After guidance from the ground, the wandering airplane is back on course toward home. Thousands of flight assists are reported each year by FAA personnel, and the true number is probably much higher because help is also given by flight instructors on a unicom frequency.

The helpful landmarks disappear at night. If navigational needles take a suspicious turn, confusion may be moments away. There are also times while flying in the dark when you want extra reassurance that the aircraft is, indeed, located on the chart where your finger points. Perhaps a mechanical problem is prodding you to take a fast heading to the nearest airport without calculating a new course. An excellent service in time of such stress is the DF (Direction Finding) steer.

Sprinkled throughout the country are air-traffic facilities equipped for VHF direction-finding. Unlike radar, the DF steer doesn't bounce signals from the aircraft skin or trigger a transponder. A DF station captures the signal transmitted by a plane's conventional VHF radio and converts it to bearing information. The ground operator sees the aircraft displayed as a stroke of light on a TV-like screen. With this information he can give a pilot magnetic headings to fly. In an emergency the DF steer can direct you to an airport for landing. If you're concerned only about where you are, it can fix your position by a cross reference from a VOR or a second DF facility. If weather is bad, a DF operator can announce turns and descents to get you safely on the ground.

FAA encourages pilots to try a DF steer purely for practice. It's excellent preparation for the pilot, and also enables the controller to meet his proficiency requirements of several steers each month. DF facilities can be checked in AIM, and they also appear on aeronautical charts. If you don't have this information handy during flight, a request from the nearest Flight Service Station quickly provides the DF assistance you need.

Surveillance radar

Nearly half of all flight assists by FAA are through DF steers. Many large airports also operate radar equipment which can bring a lost pilot to within a mile of a runway in a surveillance approach. The plane needs no navigational system other than a conventional communications radio. The system is intended to assist instrument traffic, but the nighttime VFR pilot can exploit it in an urgent situation.

During the first phase of a surveillance approach the controller announces headings which bring the airplane onto an extended centerline of the runway. Any headings or altitudes you hear are repeated back to the controller to avoid any misunderstanding. As you line up on final leg, the controller announces your distance from the runway at one-mile intervals. Altitudes are announced, along with small heading corrections, to keep you from wandering off course. One mile from the approach end of the runway the radar operator terminates his assist-

75. DF (Direction-Finding) antenna enables controller on ground to guide lost pilot solely by two-way radio.

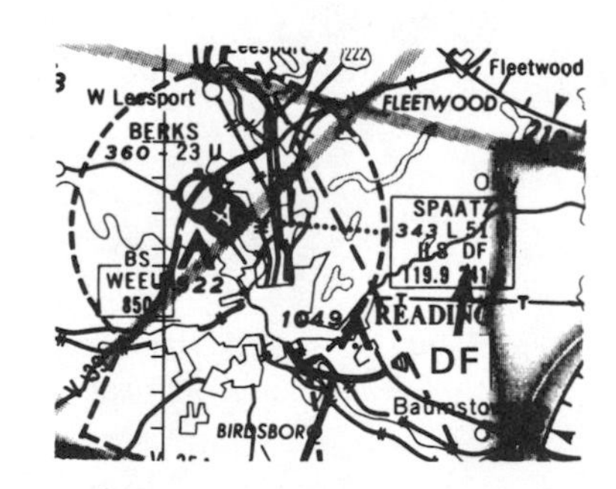

(A) DF facility shown in airport information box on sectional chart.

WACO MUNI (ACT) ***IFR*** 6NW
516 H66/18–36(3) (S–26, D–38)[1] BL5,6,8 S5
Rwy 18
Remarks: [1]Rwy 14–32 (S–57, D–68, DT–104)
Waco Tower 119.9 122.5R
App Con 119.3 122.5R 115.3T 125.9[1]
Dep Con 125.4
VHF/DF Ctc twr.

(B) DF facility shown in AIM airport directory. "Ctc twr" means "Contact tower."

76. Facilities for obtaining DF steer.

77. Terminal radar operator can instruct lost pilot.

ance, because you should have the runway environment in sight and be in a position to land visually. When traffic is light some evening, request a practice surveillance approach. The radar frequency is listed in AIM.

Radio failure

"Drop a box of chaff every two miles until four drops have been made . . ." is an official recommendation of what to do when the radio fails. Chaff, a confetti-like aluminum foil, floats through the air and enhances your target on a radar screen. It informs the radar operator that your aircraft is suffering communications failure. If chaff isn't aboard, another official suggestion is to fly the airplane in a triangular pattern to the right, using two-minute legs. Turning to the right tells the radar operator that only the transmitter is out and the receiver still works. He can transmit instructions in the blind and expect you to hear them on 121.5 MHz, the emergency frequency. If transmitter *and* receiver are lost, the procedure is the triangular pattern with turns to the left. The alert operator may respond by sending a rescue aircraft aloft to escort you back to earth.

These radio-failure techniques—dropping chaff and flying triangles—have been put to actual tests by private pilots. Both were dismal failures. (When chaff was released from the window of one light plane, hundreds of aluminum slivers blew back into the cockpit and showered the pilots!) In one trial over the radar-saturated coast of California, pilots flew hours of triangular courses and failed to attract attention from the ground. Many airmen would probably not consume their remaining fuel on triangular patterns or waiting for help. There's far more sense in other FAA advice about radio failure: When it occurs in VFR conditions, land as soon as practicable. When the flight is IFR, radio failure is serious, and the rules spell out precise routes and altitudes to fly.

The transponder, aboard tens of thousands of single-engine aircraft which fly VFR, provides another link to the ground. If you are unable to transmit and need to communicate, "squawk" the appropriate radio-failure code on the transponder and listen to the receiver on 121.5 MHz. The controller may ask a question, then expect you to reply by pressing the transponder's ident button. He sees a blip on his screen. If your condition threatens to deteriorate beyond simple radio failure, set the transponder to the *emergency* code to command the most urgent attention from air traffic control.

Most cases of radio failure are minor because many aircraft have dual navcoms and ADF receivers. The loss of one radio during VFR flight becomes a mere inconvenience. When the trouble is electrical failure, however, all radios ultimately expire and may seriously affect the safety of further flight in the dark. Since this is covered in great detail in Chapter 9, consider here ways to cope with radio troubles.

Repair in the air

Much of the cost of a well-equipped light plane is in its electronic equipment, but a bit of torn paper in a loudspeaker or a broken microphone cable can disable communications. For this reason, a cautious night flier carries two hedges against in-flight radio failure: a spare headset and a microphone. Either item is quickly jacked in for the repair.

Other troubles can be cured by an impromptu "fix" while in flight, because many radio ailments are not electronic but mechanical and superficial. A good example is the frequency selector on navcom radios. Contacts inside the switch become coated with dirt and make poor electrical contact. You discover it when a ground station fails to answer your transmission or a navigational needle wavers. In some cases, only certain channels become inactive or erratic. Try to restore a dead channel by rocking the selector back and forth several times to temporarily clear the internal contact. Another trick is to grasp the selector switch and apply slight pressure in all directions. You may strike a position that reactivates the speaker or needle.

Another troublesome set of contacts is in send/receive relays which control VHF transceivers. Dirt on their surfaces weakens or interrupts radio signals. Verify it by keying the mike button quickly several times; it may clear up the trouble.

Cables are notorious troublemakers that also respond to first aid. If you hear reports of intermittent transmission, suspect the mike and its cable. Some cases are cured by vigorously rapping the mike against the palm of your hand. One weak point is where the mike cable attaches to its plug. Try pushing, twisting or flexing the cable at this point to restore the break. (If successful, you'll have to maintain this position.) Speakers that cause badly distorted reception are sometimes fixed by gentle finger pressure through the overhead grille. In one case, a dead microphone was brought to life by jiggling a loose mike-selector switch on the audio panel.

A classic cause of radio failure sounds unlikely but it happens regularly. It's forgetting to place audio-selector switches in the correct position. This silences the loudspeaker or disables the microphone and the radio is apparently dead. Setting a squelch control too high can also completely kill a receiver, so turn the control until the static *just* disappears.

Emergency homing

When radionavigation and communications fail, all is not lost. Nearly any transistor portable radio is a primitive direction-finding instrument. You've probably noticed that best reception on AM broadcast stations occurs when the radio is rotated for loudest sound. Inside the case a loop-stick antenna responds most strongly when it lies broadside to the radio station. When the end of the loopstick points to the arriving signal, the receiver is *nulled* and the program may completely drop out. A few moments' experimentation reveals that one complete rotation of the set produces two nulls 180 degrees apart. Skillfully handled, the null point may lead you to a radio station or airport.

The crudest homer is a small AM portable operating in the standard broadcast band. You can investigate its homing possibilities on a tabletop at home. Place a local sectional chart on the table and orient its longitude lines so they actually run true north and south. A small magnetic compass is helpful here. Tune in a local radio station of known location and rotate the whole radio until the audio drops out at a null. Next, verify the directionality of the radio; that is, note which side of the case lies along a line of position toward the station. Take the radio on a flight and try a few test runs to see if you can home toward a station with reasonable accuracy. Hold the radio level and line up the side of the radio case (that gives the line of position) with the fore-and-aft centerline of the airplane. Turn the airplane until a null is heard and note the compass reading. This heading is a line of position toward *or* away from the station. Ignore the ambiguity for a moment and assume you know the correct null to follow. By steering the compass course, while keeping the signal *nulled* (for lowest audio), you should eventually reach the station.

Portable radios manufactured specifically for direction-finding make the job easier. The loopstick mounts at the top of the radio on a pivot. Instead of rotating the radio, only the loop is turned until the null is heard. These radios have calibration marks which indicate how many degrees the station lies to the left or right of the aircraft's nose; this is especially helpful if you want to triangulate. Or tune two or more stations and draw their lines of position on your chart; where the lines intersect is your location. This is one way to resolve the ambiguity caused by two nulls when the antenna or radio is turned over a full circle. In most instances, however, you know your approximate position and it is obvious that one null is utterly wrong.

ADF portables also have a tuning meter which rises and falls to reveal the null points as the loop rotates. However, it's usually easier to hear the change in audio loudness during a null rather than seeing the

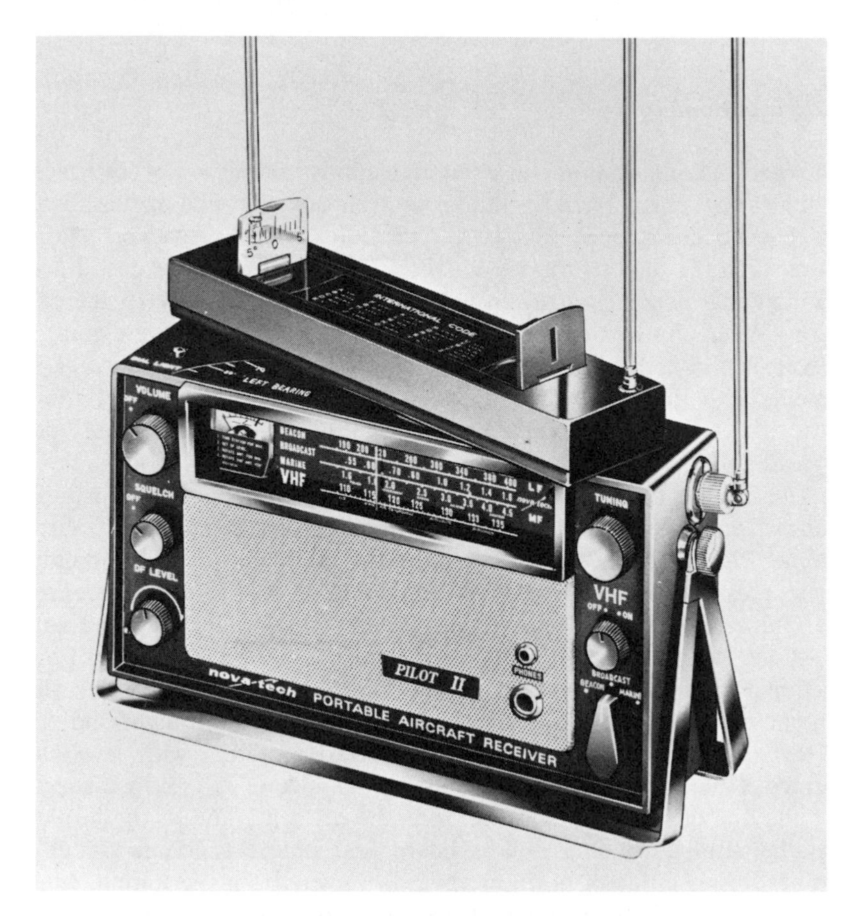

78. Portable direction-finding receiver.

movement of the tiny needle. Some sets have rotatable azimuth dials that display headings and intercepts without requiring the pilot to do a lot of arithmetic.

If you plan to use a small portable as an emergency backup, take it on a shakedown cruise when the weather is CAVU (Ceiling And Visibility Unlimited). One or more of the following pitfalls might show up, and a clear day is the time to deal with them:

Which way? Observe the 180-degree ambiguity and practice how to follow the correct null. Triangulation was mentioned above, but fumbling with charts, pencil lines, holding the radio and swinging the loop won't be easy when you're lost. If you know your position at the time of radio failure, selecting the correct null is simple logic.

Build-and-fade is a classic method for telling whether you're on the right track. If a station grows stronger or weaker with time, you're flying toward or away from it. A rapid and tremendous signal buildup means you're in the station's immediate vicinity. Audio becomes garbled when the plane is over the transmitting site.

Choice of station. Homing on standard broadcast can be uncertain. The greatest deficiency is the infrequent station identification. When the announcer finally utters call letters and location, you may assume that the transmitter (the point you're homing on) is actually located several miles outside of town.

Most such problems are solved by using a portable equipped with a low-frequency radio-beacon band (from approximately 200 kHz to 400 kHz). Station identification is continuous and positive, and stations are listed on aeronautical charts.

Audio volume. Do not rely on the loudspeaker in a small portable to deliver adequate audio volume for tuning a station, hearing its identifier or nulling. Engine noise in a light plane overwhelms the speaker, so carry an earphone for listening.

Ignition noise. Airplane spark plugs, alternator and other electrical devices create considerable noise in a portable radio. Depending on the degree of noise suppression, some aircraft are worse than others. During flight tests you will hear that strong stations silence the interference and permit adequate direction-finding. The noise problem grows worse as you tune in lower frequencies.

Portable transmitters

The portable receiver idea can be carried a step further. Some pilots fly with portable transceivers that can send and receive on standard aircraft frequencies in an emergency. Powered by their own batteries, the sets can transmit several dozen miles from inside the cabin on a built-in telescoping antenna. Range is shorter than that of a standard aircraft radio, but often sufficient to contact an air-traffic facility on the emergency frequency.

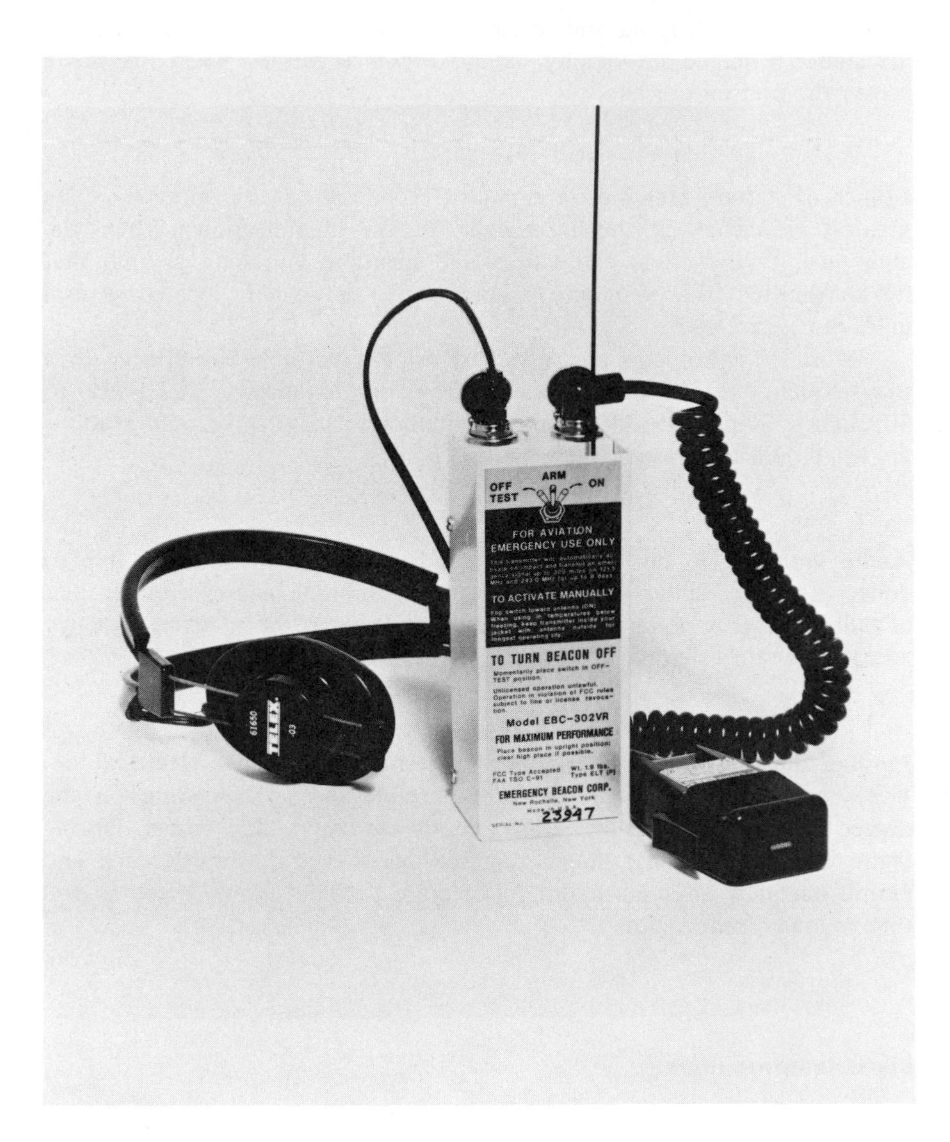

79. **Emergency locator beacon with two-way voice communications.**

Another backup system is the ELT, or Emergency Locator Transmitter, which became mandatory for most planes in the early 1970s. Some models are fitted with a jack to receive the aircraft microphone and can transmit voice on 121.5 MHz. Do not expect great transmitting range from an ELT. It emits special tone signals for hundreds of miles, but voice transmission may be limited by the sensitivity at the receiving end. The plane should be in the vicinity of a known ground station (a control tower, for example) to assure reception. Some ELT models have a small receiver to form a complete, emergency backup transceiver (Fig. 79).

Cross-country after dark

8

Night approaches an airport as the sun dips behind a hangar and lengthens the shadows across a runway. But legally speaking, "night" is not a nebulous period between dark and light. It happens at a precise moment inscribed in the *American Air Almanac.* A one-minute difference can change the responsible party in a court proceeding, create a violation, or influence an accident investigation. Before making a cross-country night flight, a pilot should know about three distinct times that affect him when the sun goes down.

If passengers are carried, a pilot's recent experience requirement (three take-offs and landings at night in the previous ninety days) applies from one hour *after* sunset until one hour *before* sunrise. Turning on the aircraft's position lights, on the other hand, starts *at* sunset. Lights must be illuminated whenever the plane operates between that time and sunrise. The third time is *night;* officially the period between the end of evening civil twilight and the beginning of morning civil twilight.

Two cases show how meticulously the regulations may apply. A Piper Colt with no position lights was reported flying in the vicinity of an airport at 7 P.M. one evening. Although the pilot didn't know it was night, sunset had officially occurred at 6:56 P.M. The four-minute difference was sufficient to start an NTSB investigation. Since there was no other evidence of reckless or dangerous operation, the pilot, thanks to an act of official mercy, was not punished.

In the other case, the pilot wasn't so fortunate. During an approach to a Las Vegas airport at 6:56 P.M., an aircraft struck power lines and crashed, killing the pilot and seriously injuring two passengers. Investigators determined that the pilot was operating beyond his ability at the time of the accident because his medical certificate bore a restriction against night flying. Night was officially fixed at seventeen minutes before the accident.

With reasonable judgment, a pilot usually does not have to determine the official times with great precision. Position lights can be turned on well before sunset—the time when the top of the sun's disk exactly touches the horizon. If you want to verify an official time, simply ask a Flight Service Station, control tower or weather bureau. They have appropriate schedules for the local area. Daily newspapers often list the time of sunset and sunrise.

Another valuable item in the local newspaper are the phases of the moon. Note that August 27 in Figure 81 marks a new moon, meaning it will be a dark night. Since the accident rate is far greater on moonless nights than on bright nights, a pilot with little night experience may wish to avoid this time on his first cross-country. Flights under a bright moon afford a gentler transition. Bodies of water glisten and create excellent checkpoints (Fig. 82).

Sun and Moon

The sun rises today at 6:13 A.M.; sets at 7:44 P.M.; and and will rise tomorrow at 6:14 A.M.
The moon rises tomorrow at 12:05 A.M.; sets at 4:27 P.M.; and will rise Friday at 1:58 A.M.

81. Daily newspaper may give local sunrise, sunset and moon phases.

82. Bright, moonlit night makes water highly visible.

Preflight tips

Since basic flight planning is amply covered in other publications, consider here certain sensitive areas of interest to a night flier:

In a preflight weather briefing, carefully check the temperature-dew-point spread. When these numbers merge to within a few degrees of each other it's a powerful warning that visibility may soon drop. Un forecast weather is another danger on a long cross-country at night, so monitor conditions by radio while en route and be ready to divert to an alternate airport. Another significant weather item is the winds-aloft forecast, since higher altitudes are often flown at night, where winds are considerably stronger than on the ground. Unless you've allowed for headwinds, they can seriously eat into a fuel supply and reduce the aircraft's range. The danger of low fuel multiplies at night because you need a lighted airport for landing, and one is not always available. Also, few airports are open after dark to refuel an aircraft. Know where large, all-night airports are located if an in-flight diversion for fuel becomes urgent. Plenty of reserve and careful estimates should prevent it.

File a flight plan at night. If you're departing from a small airport with no phone to a Flight Service Station, call in your flight plan from home or office. You can do it hours in advance of the flight, then activate it in the air. Besides the flight plan, give voluntary position reports by radio while en route, and ETAs to your next checkpoint. The safety factor is obvious, and position reports keep you aware of your course and position.

En route

Let no radio in the airplane remain idle. With dual VORs, don't merely track a radial; choose an intersection on the other set to freshen your position along the way. Tune the ADF receiver to a beacon or broadcast station as a backup. If any navigational radio suggests you're drifting off course, don't delay making a correction. Night winds quickly blow a small plane off track, and returning to course may take considerable time.

Identify airport beacons as they pass in the night, even if they are not on your course. If trouble develops, you can point the nose at an airport beacon and land in minutes without a lot of knee-board navigation.

Keep a transmitter on a frequency that can bring immediate help. Unicom is a poor choice unless you are near an uncontrolled field, with no tower, announcing take-off or landing intentions. While en route, switch to a control tower or Flight Service Station. If two radios are in the plane, set one to 121.5 MHz, the almost universally monitored emergency channel.

83. Radar assistance for the VFR pilot is available at many airports.

The sophisticated VFR night flier communicates with radar operators through much of a cross-country flight. When work load permits (and it usually does at night), controllers obligingly track your progress on radar screens and advise on other traffic. They can give vectors to an airport and render immediate assistance if you develop a problem. Frequencies of these radar services, listed in AIM, can be recorded in your flight log, or call a Flight Service Station and request the name of the nearest approach, departure or "center" radar.

Arrival

A boon to a night flier traveling to busy airports is Terminal Radar Service. A VFR flight operating through such airports may be handled much as if it were on an IFR flight plan; the pilot receives landing information, vectors, sequencing and separation from other traffic. Only a two-way radio is needed to participate, and neither transponder nor instrument rating is required. To discover if this service is offered, check Part 3 and Part 4 of AIM. It is indicated in the airport/facility section as TRSA, for Terminal Radar Service Area. A chart may also appear in Part 4 showing the airspace, landmarks and terrain features for a radius of about fifteen to thirty miles around the airport. Although the service is rendered during VFR conditions, a night arrival and departure are much safer under radar surveillance.

Arriving at an uncontrolled airport is another matter. AIM may describe lighting facilities and obstructions, but an arrival in the dark is often greeted by silence on the radio. To minimize the chance of conflict with other aircraft flying near your airport, the FAA recommends that you make one-way radio announcements while arriving or departing from almost any type of airport. As shown in Figure 84, tune your radio to the appropriate frequency before your arrival, pick up the microphone and make the first broadcast at least five miles from the airport. Tell where you are, your altitude and intentions: e.g. "Cessna, five miles northeast Smithtown Airport at 3,000 feet, landing." This warns other aircraft in the area, which should be monitoring the frequency. As you approach, you'll hear the position and intentions of other aircraft following the same procedure.[1] Notice that if the airport has an FSS in operation, make the first call fifteen (not five) miles away. Safety is also enhanced by following similar procedures described in the chart of Figure 84 as you taxi and take off for departure.

[1] A pilot may feel that no one is listening while he broadcasts, or is "talking in the blind." Not always. One night the author was on final approach to an uncontrolled feeder airport and spotted an airliner on the taxiway. The captain warned other aircraft that he was about to take off by asking, on the unicom frequency, "Let's see, which way is New York?"

INBOUND AIRCRAFT

Facility	Frequency	*First Broadcast	Repeated Broadcasts
Part-time Tower (when closed)	Tower Local Control	5 miles	Downwind, Base, Final
Part-time Tower (closed) but Full-time FSS	Tower Local Control	15 miles	5 miles
Part-time FSS (closed)	123.6	5 miles	Downwind, Base, Final
Full-time or Part-time FSS (open)	123.6	15 miles	5 miles
UNICOM	122.8	5 miles	
UNICOM (if unable establish contact)	122.8	5 miles	Downwind, Base, Final
No Facility on Airport	122.9	5 milcs	Downwind, Base, Final

OUTBOUND AIRCRAFT.

Airport	Frequency	Broadcast Position and Intentions
Part-time Tower (closed)	Tower Local Control	At all uncontrolled airports (regardless of facility) departing aircraft should broadcast position and intentions when ready to taxi and before taking runway for takeoff.
Part-time Tower (closed) but Full-time FSS	Tower Local Control	
Part-time FSS (closed)	123.6	
Full-time or Part-time FSS (open)	123.6	
UNICOM (whether or not responding)	122.8	
No Facility on Airport	122.9	

*On first broadcast announce position, altitude and intentions.

84. When tower or unicom are closed down for the night, a pilot can broadcast one-way announcements on the radio to make his presence known to other aircraft in the vicinity. For example: in the first case (see "Part-time Tower when closed"), tune radio to tower frequency and announce your position when five miles from the airport. Also broadcast position when entering each leg of the traffic pattern.

A positive procedure at a strange airport is to circle to the left over the field to study the windsock. As mentioned in an earlier chapter, a wind tee or tetrahedron could be tied down for the night and not be responsive to wind direction. (And a downwind landing at night is risky.) Use the lighted windsock to determine the wind, then make blind announcements of your position and intentions in the pattern. Sometimes you're greeted on the radio by a local pilot with helpful information about the airport and landing direction.

Touchdown and roll-out don't mark the end of a safe cross-country flight. There is the trip to the ramp or tie-down. A ground controller takes care of you at a large airport, but in the boondocks taxiing may be the most indecisive time of the trip. Many small fields have no taxiway lights or line crews to guide you in the evening. Cautious taxiing at low speed, while studying the area illuminated by the landing light, might be enough if the light strikes small ditches, chocks lying about or other obstacles. If there's any uncertainty about a darkened area, stop the airplane, get out and scan the area on foot with a flashlight. Do not allow a passenger, who may have no experience with aircraft, to direct you by walking ahead of your moving airplane. The spinning prop is too dangerous to someone unfamiliar with an aircraft maneuvering on the ground.

Instrument proficiency for night cross-country

Once you've completed your first cross-country at night, and the aircraft is securely tied down, you'll walk away with a feeling that's touched thousands of pilots before you: "That wasn't so difficult, after all." The aircraft performed faultlessly, the engine never missing a beat over twinkling cities and inky blackness. You'll agree that meticulous flight planning took most of the surprise out of flying in the dark. With good visibility and clear skies, you discovered that flying at night is almost as effortless as flying during the day.

But it takes more than a brief cross-country to become a competent night pilot—and roar off into the dark anytime the sequence reports tick out VFR for every terminal within two hundred miles. Even under favorable conditions you'll need basic instrument proficiency before attempting any extensive cross-country flying. These aren't the advanced maneuvers and procedures demanded of an instrument-rated pilot, but currency in the rudimentary skills of a private pilot. The FAA requires that all pilots be able to maintain control, day or night, when outside references are lost.

These skills could be tested more than once on a typical night cross-country. The daytime flier enjoys a continuing spectacle of highways, farms, rivers and towns revolving below in a fascinating panorama. These features, too, provide a pilot with attitude information needed to keep the craft tracking a normal flight path. A trip in daylight, in fact, can be done without ever glancing at the instrument panel.

Compare this with a long flight over a region with few ground lights and a dark sky. After an hour the drone of the engine and lack of apparent motion outside the cockpit isolate pilot and plane from the earth below. As distant points of light creep toward you from over the horizon they form surrealistic patterns which reveal nothing about the contours below or how they relate to a plane's attitude. Somewhere into the flight, a look at the panel may reveal that something's amiss: a wing that's eased into a bank, a decay in airspeed because the nose drifted upward, or an altimeter that's slowly unwinding. During the day the eye's uncanny sensitivity catches these cues from without. At night a slight change may slowly, but inexorably, develop into a dangerous attitude. When a seat-of-the-pants warning finally jolts the pilot, the airplane may already be in a threatening condition. Because these situations arise from sensory deprivation at night, a pilot must come to rely on, and trust, his instruments.

It is not *total* dependency. Any night flight which requires continuous assistance from the aircraft gauges should be done only by an instrument-rated pilot. The instruments during good VFR, on the other hand, can *verify* that all is going well. When you leave a shoreline and enter the blackness of overwater flight, it's reassuring to glance at the artificial or gyro horizon to be sure the wings are level. Bank sharply over a dazzling carpet of city lights while flying the pattern around an airport and you'll want to peek at the tiny wings on the artificial horizon to be sure they don't tip beyond a gentle bank. When taking off from a field surrounded by high terrain and you pitch up the nose for good climb, the artificial horizon gives instant reassurance that your ascent won't skewer upwards into a deadly stall.

An excellent way to use instruments on a cross-country is to include them in an overall scan of the eye that sweeps a wide arc outside the cabin, then dwells briefly on each instrument. Such an all-encompassing scan will almost certainly catch a minor deviation that needs correction.

The outside-inside scan adds another benefit to night flying. It prevents fixation, the tendency to stare hypnotically in the dark or gaze at a single instrument. Too much time spent looking at one spot triggers the optical illusions described in an earlier chapter or allows the plane to go awry in some other dimension. Scanning isn't done with robotlike repetition, because that, too, dulls a pilot's responses. Don't simply satisfy

85. Useful instruments for verifying attitude.

a compulsion to drive your eye over each gauge, but develop an awareness of what each needle is saying. Follow the scan by small corrective control action when needed.

If your plane is equipped with a typical instrument panel found (Fig. 85) in many light aircraft, it includes a gyro (artificial) horizon, an air-speed indicator, vertical speed indicator (for rate of climb), a turn-and-slip indicator (or "turn co-ordinator") and an altimeter. It's possible to fly reasonably well with only two of those gauges in the time-honored technique that launched instrument flying: "needle, ball and airspeed." Since the turn needle indicates a rate of turn, keeping the needle centered with aileron control keeps the wings level. The "ball" moves from its central position whenever there is a turn of poor quality and the plane is skidding or slipping. Both needle and ball are kept centered for straight and level flight by co-ordinated movements of aileron and rudder. Mean-while, a steady reading on the airspeed indicator keeps the nose level along the pitch axis.

Thus it's possible to monitor a night flight with only needle, ball and airspeed. A glance at a left-leaning turn needle, for example, suggests that the left wing is low and needs lifting by the right aileron. If the ball slides to the right during an intentional turn (in the traffic pattern, for example) you'll avoid a potentially dangerous skid by centering the ball with the rudder. Since dangerous attitudes may develop quickly at night and require fast pilot response, it's a good idea to memorize this rule of thumb: If the ball is off-center (indicating a sloppy maneuver), *step on the ball* to correct the plane's attitude. This simply means step on the rudder pedal on the same side as the ball. At the same time, the turn needle is centered to restore a wings-level attitude by operating the aileron.

If your aircraft has an artificial horizon, consider it the primary attitude indicator for night flight. It presents the most graphic, centralized image of the plane's nose and wings and is remarkably easy to understand. At a glance you receive pitch and bank information from the miniature, symbolic airplane for effortless interpretation. There is, however, a possible pitfall in using an older-type artificial horizon (where the miniature airplane is fixed with respect to the instrument panel, while the horizon tilts). When the real plane is in a well-developed bank, and the wings of the miniature plane are also tilted, it is sometimes difficult to quickly tell which way the plane is turning. This could be critical when you're rapidly running out of altitude. If there's any confusion about which way to turn to make a recovery, glance at the turn needle. It unerringly leans toward the low wing and lucidly tells which way to move the aileron to restore level flight.

Also check airspeed and vertical-speed indicators to verify the aircraft pitch. An artificial horizon may indicate a nose-up condition, but this is no proof that the plane is actually climbing. Those other instruments verify it as airspeed drops off and vertical speed increases.

True instrument flying is beyond the scope of this book and is amply covered in other publications. But an ability to read and interpret the aircraft's attitude instruments in the most basic fashion is an inextricable part of safe night flight. If you're rusty on the gauges, take a few hours under the hood with an instructor or safety pilot. Night flying doesn't demand that you fly blind for hours, climb and descend through cloud layers, then break through an overcast several hundred feet above the ground and land on a wet runway. All a night flier needs to receive from his gauges is reassurance that darkness isn't luring his aircraft into a potentially unsafe attitude or maneuver.

CIRCUIT BREAKERS

Electrical failure

9

Famed aircraft designer Jim Bede suffered complete electrical failure one night while attempting to break a record for long-distance flight. Although he was trapped in darkness above a cloud deck, there was little cause for alarm. A rescue plane was sent aloft to lead him down through the layers to a safe landing. Bede later dismissed the in-flight emergency as "a lot of fun." The average night flier, unfortunately, has no such standby service.

Sudden and total blackout rarely happens. It's well known that an airplane has dual magnetos and two spark plugs per cylinder to prevent complete engine stoppage if the ignition fails. Not so familiar are its two electrical systems: a storage battery, and a charging system equipped with a generator or alternator. Either one may power a plane's electrical system, and simultaneous loss of both would be an extraordinary event. According to battery manufacturers, the leading causes of battery failure are overcharging, undercharging and lack of water. But none of these causes should extinguish a battery in minutes. A good sign the battery is still vital is that it started the engine on the ground—or you shouldn't have attempted a night flight in the first place!

When catastrophic electrical failure occurs in flight, it often begins in the alternator or generator. A component burns out, cable connections loosen, a voltage regulator fails, or a drive belt breaks. As the charging system dies, however, instruments continue to indicate and lights shine because the electrical load instantly diverts to the battery. Up to the moment of failure, the battery merely "floats" on the charging system, absorbing a trickle of an ampere or two. If a generator suddenly drops off the line, battery voltage immediately drives current into the breach.

The changeover from charging system to battery power is so subtle it easily escapes the pilot's eye. This is an advantage, on the one hand, because the battery is automatically an emergency backup. But the smooth takeover also carries a disadvantage: Enough life may secretly ebb from the battery to turn a situation of mild concern into an emergency. When the battery surrenders its last ampere it disables a considerable array of equipment: cockpit illumination, navigation, anti-collision and landing lights, communications and radionavigation, engine instruments (fuel gauges drop to zero), electric fuel pump, stall warner, landing gear and autopilot.

The loss is a minor inconvenience on a local flight under good VFR conditions—grab a flashlight and fly toward the airport beacon. But on a cross-country the situation grows grave unless a pilot can navigate accurately by dead reckoning to a lighted airport. An idea of what it's like to lose both battery and generator systems is experienced by turning off the master switch with the engine running while the plane is on the ground. On a dark night, even as a flashlight picks out the instruments, it is a foreboding experience.

None of this is necessary if the battery is never permitted to completely expire. There are techniques to eke amperes out of a flagging battery from the time the charging system stops until touchdown. A good starting point is to understand the most important tool for fighting electrical failure.

How to read an ammeter

Electrical systems on early aircraft were little more than whirling magnetos to fire the spark plugs. As lights, radios and other accessories were installed, the magneto was hardly up to the task of supplying additional power. A magneto usually delivers high voltage at tiny currents, and most accessories demand the opposite combination. Engineers adopted the automobile charging system and transplanted much of it to aircraft. (A close look at an aircraft generator, alternator or voltage regulator reveals a kinship to the automobile version.) Presiding over the ebb and flow of currents is the ammeter, whose subtle movements reveal the state of the electrical power system.

The generating system and battery form a self-contained power plant that develops nearly 1,000 watts. In a typical light plane it is controlled by a master switch that is isolated from engine ignition. This is an important consideration because it shows the independence between magneto and generating systems. Because of this separation, you can turn off a master switch while in flight without affecting engine operation. As we'll see, this is a critical step in solving certain emergencies, so a night flier should verify, at least once, what actually happens when the master is switched off. During a run-up on the ground, experimentally turn off the master to check whether lights, radios and other electrically driven instruments expire—then return to life when the master is turned on again.

As the master switch is turned on and off several times, notice the vigorous action of the ammeter needle as it senses and measures current flow. Your aircraft may have one of two ammeter styles, so consider these models in some detail. Each reads a different set of electrical conditions.

87. Zero-center ammeter.

Zero-center ammeter

The earlier of the two types, the zero-center ammeter, is almost always found in systems powered by a generator. When the needle deflects to the left—to the negative, or discharge, side—the ammeter indicates a flow of current *from* the battery. It's axiomatic that continuous, unrelieved discharge will cause a dead battery. How long it takes depends on the current draw and the battery's rated capacity. For example, a battery for some light aircraft is rated at 25-ampere hours, meaning that, fully charged, it can deliver almost 25 amperes for one hour. If the discharge rate is faster, efficiency drops and the battery delivers less than full capacity: e.g. 140 amperes for five minutes or 48 amperes for twenty minutes. These battery drains are reflected in practical ammeter readings.

When the pilot boards the aircraft, the ammeter indicates zero (the drain of a plane's electric clock is too small to register). As the master is turned on, any electrical instrument that has no on-off switch—turn-and-bank indicator, flight recorder, etc.—indicates a discharge on the meter if it's more than an ampere or two. (You may also see the draw of radios or lights you forgot to switch off after the previous flight.) Any electrical accessory turned on at this time causes the ammeter to read the total number of amperes discharged by the battery.

Next, the starter is pressed. As the engine turns over, the starter may draw more than 100 amperes from the battery. This will not show on the ammeter. To protect the meter's delicate movement, heavy starter current is diverted around the ammeter. When the engine comes to life, it spins the generator to convert about one or two horsepower into electrical energy. Now the needle swings out of the discharge region and into the positive, or charging, segment of the ammeter. If the engine is held at a fast idle, the ammeter needle indicates a high rate of charge for several minutes as the generator replaces energy lost by the battery while turning the starter motor. Depending on the aircraft, generator output for light planes is in the 25- to 60-ampere class.

As the airplane taxis toward take-off during the day, the ammeter makes darting motions in the downward direction away from the high-charge region. It means the voltage regulator has sensed that the battery is again fully charged and is automatically reducing generator current. At night, however, with landing lights and other electrical equipment in operation, the ammeter usually remains deeply in the discharge region. Low engine rpm prevents the generator from delivering its full output current, so the battery contributes enough amperes to support part of the electrical load. When the engine speeds up for take-off, a swing of the ammeter high into the charge region shows that the generator now powers all electrical items and is replenishing the battery.

About twenty minutes after take-off, in cruising flight, the zero-center ammeter should remain at *approximately one-half needle width to the right* of zero. This is the normal cruise condition of the electrical system; the generator is imparting a slight trickle charge to the battery and powering all active electrical accessories. That last point is important. If you turn on additional radios or lights in cruising flight, the ammeter needle should *not* significantly change. It may flick downward momentarily but should always return to its basic position—slightly in the charge area. A constant reading, even as you switch accessories on or off, means that the aircraft's electrical system is correctly regulating the generator flow to the load. As shown in Figure 88, the ammeter doesn't

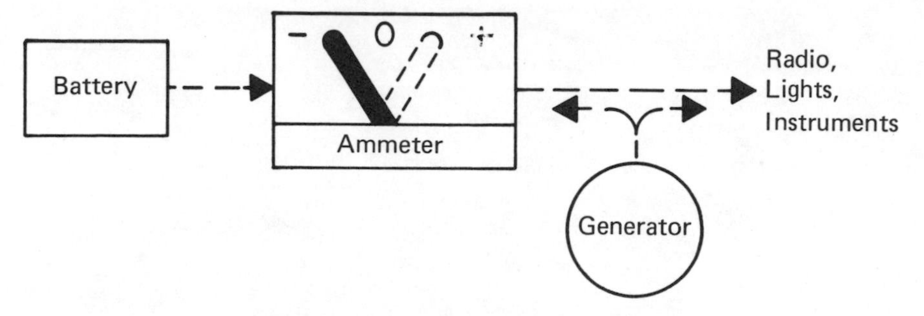

88. Position of zero-center ammeter in system using a generator.

monitor current drawn by accessories from the generator, because of the meter's position in the electrical system.

If you suspect electrical trouble in flight, make these quick checks. Turn on any heavy-current device—a landing light or pitot heater—and watch the ammeter. The needle may gyrate for a moment as the regulator readjusts to the new load, but it should soon settle down to the steady trickle-charge position. If you turn on every electrical switch in the aircraft, and the ammeter drops to discharge (at cruise rpm), the electrical system may be inadequate or faulty. Regulations prohibit the installation of electrical equipment in an aircraft unless the charging system can support it.

There are minor exceptions to this rule. At night, when the electrical system is operating close to full capacity, the ammeter shows deep discharge during a gliding descent or while taxiing. At low engine rpm, the generator rotates too slowly to deliver full output, and the battery supplies the missing current. This is a temporary condition and the battery should yield sufficient reserve to carry the electrical system for brief periods. Another slight anomaly is that an accessory which operates in pulsating fashion—a beacon or strobe light, for example—may cause rhythmic flickering on the ammeter needle.

Remember the most important rule for a zero-center ammeter: After twenty minutes of cruising flight it should indicate an ampere or two—or one-half needle width—of charge. Steady discharge, even of small proportions, is cause for alarm and immediate action at night.

Left-zero ammeter

During the mid-1960s, automobile manufacturers switched from generators to alternators, and airplanes joined the changeover. One advantage is that an alternator continues to deliver current at low rpm. Unlike the generator, it supplies amperes as an airplane taxis or the engine turns at less than cruise speed. At the time of the alternator changeover, the ammeter also received a new design. Instead of a zero-center needle, the pointer rests on the zero point at the left, much like the needle of a fuel gauge when the tank is empty.

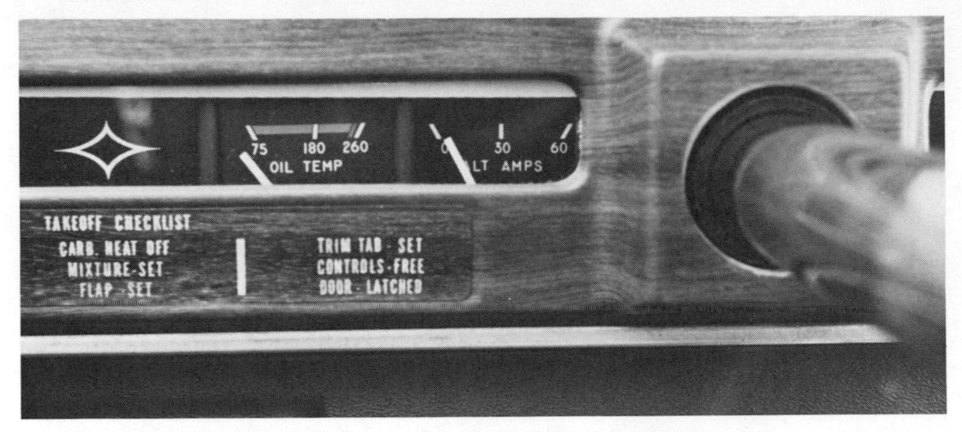

89. Ammeter in alternator system is seen at right.

The new ammeter ignores the battery flow and indicates the *total* amount of amperes drawn only from the alternator. As each accessory is switched on, the ammeter climbs higher on the scale. This changeover has caused confusion among pilots who associate a rising ammeter reading with a higher rate of battery charge. This is not the case in this ammeter type, and the simplified illustration in Figure 90 shows why. Instead of a position between battery and generator, the ammeter is now inserted just *after* the alternator to read only alternator output. Now when the pilot enters the airplane and turns on the master switch, no current is indicated on the ammeter (the alternator is not turning). If a radio or light is turned on, again nothing registers on the ammeter because battery discharge current is not shown on this ammeter type.

Next, start the engine. As the alternator develops power it delivers charging current to the battery. This appears on the ammeter. Turn on an 8-ampere radio and the needle registers 8 more amperes. A 2-ampere panel light similarly raises the reading by that amount. Since the ammeter shows total alternator output (consumed by both battery and accessories), the reading also warns of trouble. Turning on an electrical device must show a corresponding rise in the ammeter reading. Turn off all electrical devices, and the reading is reduced to that solely drawn by the battery. If that reading ever disappears, and the ammeter reads zero in flight, the charging system has failed.

Early signs

Either ammeter type signals the earliest warning of impending electrical failure. If the pilot fails to see it, he certainly notices the next symptom. It is dimming of cockpit or navigation lights as the battery loses its remaining capacity. Since the threshold of change can be small, considerable loss of illumination may occur before it attracts the eye. Yet the

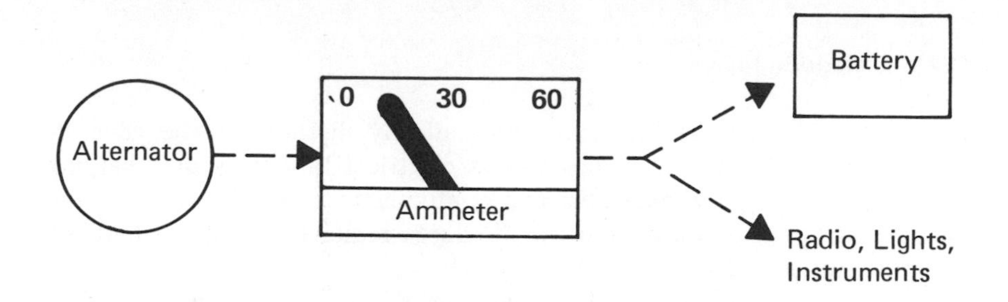

90. Position of ammeter in systems with alternator.

moment arrives when the feeble glow of the lighting system is unmistakable. A glance at the ammeter verifies it; the zero-center needle is now in the discharge region, or an alternator-type needle is resting on zero.

My first electrical failure produced another symptom. While navigating over water, the TO FROM flag of a VOR receiver grew unsteady. Since an occasional flutter is no cause for concern, the flag was ignored. Minutes later the OFF flag edged into view. It tentatively rose and fell until it completely blanked the TO-FROM indicator. This usually means that the airplane is moving beyond reception range of the VOR and it's time to select a closer station. The trouble was, I could look out of the airplane's rear window and *see* the island on the horizon where the VOR signal originated. The explanation was soon apparent. The VOR receiver was sensing a weak signal—but only because the plane's electrical supply voltage was dropping and ruining the radio's sensitivity. A look at the ammeter confirmed the trouble. Although it was barely in the discharge region, it was signaling generator failure. This was verified by snapping on the pitot heater and landing light. The needle dipped sharply to the left. (When the generator was later checked on the ground a mechanic pronounced it completely dead.)

There is reason to believe this particular failure was not instantaneous, but happened over a period of an hour or so. Despite a considerable electrical load—four radios and two lighting systems—the ammeter rested near zero and attracted no attention. During this time, the battery and weakening generator probably shared the load. Both systems continued to deteriorate (first generator, then battery) until the radios protested by dropping the VOR flags and signaled the warning.

Electrical failure can also be caused by the voltage regulator. After engine start-up, it directs a strong flow to the battery for recharging, or increases output of the alternator or generator as more equipment is switched on. The voltage regulator, however, may lag slightly behind fast changes in electrical demand and momentarily fail to control the supply voltage. This could cause a dangerous surge in the solid-state circuits of navigation and communications equipment. The threat is great enough for manufacturers to build in a protective device known as an overvoltage relay. If voltage exceeds approximately 16 volts (in the typical 12-volt system) the relay automatically disconnects the generating system from the line. It also causes electrical failure of the charging system! As the battery takes over and continues to power the aircraft, the ammeter drops to zero (or into the discharge region) to indicate that the charging system is disabled.

You must quickly determine whether the power failure is due to temporary overvoltage, because it's easily corrected. This is done by *recycling* the master. Turn the master switch off for approximately thirty seconds (or follow instructions in the airplane manual). If the overvoltage was only a transient "spike," the relay should automatically reset. When the master is turned on again, the system returns to normal and the ammeter should indicate charging current.

An overvoltage happened to one pilot who, fortunately, wasn't far from home. His electrical system was struck by a brief overvoltage which tripped the relay and shut down the electrical power. He reached for his flashlight when he noticed gradual darkening of the instrument panel. After a successful landing he was asked if he had tried anything during the flight to restore power. "No," he replied, "the engine was running and I wasn't going to touch anything that might change that fact!" Only as he taxiied to the ramp did he flick the master switch, and to his surprise, the ammeter needle swung up the scale and the lights grew bright.

The pilot was wrong in not trying to reset the relay in flight by operating the master switch. Any reluctance to turn off the master in flight should be tested, as mentioned earlier, to your satisfaction on the ground. In the standard arrangement, it can be recycled without stopping the engine.

A voltage regulator can produce another condition which appears on the ammeter. It is excessive charging rate, a malfunction that can overheat and evaporate battery electrolyte and possibly damage radios and lights. If the flight can't be quickly terminated, the strain on the system is relieved in aircraft equipped with a *split master*. In a typical system, the left half of the master inserts the battery on the line while the switch's right half controls the alternator output. In normal operation the switch is treated as one unit and interlocked so the pilot can never turn on the alternator without the battery. It's possible, however, to turn *off* the alternator but leave the battery on. This offers some control over excessive charging rates. If abnormally high charge is seen on the ammeter, *only* the alternator side of the switch is turned off. This leaves the battery on the line to deliver electrical power. Since the battery now discharges steadily into the load and may go dead, the pilot can turn on the alternator for a few minutes to manually recharge the battery. How long the alternator is energized is a matter of judgment, but dimming lights and failing radios provide ample clues. Before landing, turn on the alternator to assure a plentiful supply of power for landing lights and other electrical items.

Another use for the split master occurs in cases of alternator failure. There is a good possibility that a dead alternator will continue to consume current and impose an unnecessary drain on the battery. Switch off the alternator in these cases to extend battery life.

Fuses and breakers

A short circuit in the electrical system causes heavy discharge currents which trip the fuse or breaker. It isn't unusual, either, for these overload devices to "pop" from short-term overload or age. Pressing in a breaker to reset a circuit or replacing the fuse with a spare may restore normal operation. Be sure to *firmly* depress a breaker, because some units require considerable force to reset. If the fuse or breaker doesn't fail again after two or three trials, you may assume conditions are back to normal. To be sure, however, have the electrical system checked later on the ground.

Some alternators and generators have no visible protective devices. The 5-ampere and 60-ampere alternator breakers on one popular aircraft, for example, are completely concealed because they're self-resetting. An internal element interrupts electrical output until the short circuit is removed. At that time, the breaker automatically resumes current flow and no pilot action is required.

Finding a blown fuse or open breaker is usually not difficult, because the protected light or electrical device simply won't function. This is not always a reliable conclusion in some aircraft. A technician might have tapped several "hot" wires from the same fuse or breaker. If he combines a radio and dome light on one circuit, for example, it could lead to a potentially dangerous situation if the dome light develops a short circuit. In blowing the fuse or breaker, the light would also disable a good radio. To check this possibility, identify, on the ground, how each electrical device is protected. It's easy with fuses because a fuse is merely removed, and you see or hear what is disabled. Unfortunately, many circuit breakers cannot be turned off from the panel and the job might require the help of a mechanic or technician. Try to label the function of each breaker and fuse.

If you cannot identify an accessory that's blowing a fuse or tripping a breaker, try this procedure. First, turn off every possible electrical item. Reset any protruding breakers and replace blown fuses. Turn on one electrical item at a time. The one that blows the overload is the culprit and is left switched off. This allows you to again reset the breaker or replace the fuse to keep the good circuit operating. (Spare fuses should be aboard, but in an urgent situation pilots have "borrowed" fuses from one circuit and inserted them into another.) Powering radios, lights or instruments from common fuses is a shoddy installation practice and should be avoided.

How much power remains?

If electrical failure continues unnoticed, white lights turn yellow and red lamps turn the color of plums. In the dimming cockpit you start wondering how much current remains to see you safely home. The battery may have dozens of amperes—or not enough power to illuminate a small lamp. Now you want to skillfully manage the remaining capacity, or try to revive a dead battery for a final gasp to call for help.

When the battery is fully charged—which it should be during normal flight—it is rated to deliver about 30 amperes for one hour in a typical light plane. That figure is also the approximate number of amperes continuously consumed by lights and instruments during night flight. Thus, at the moment of failure you might anticipate emergency power for up to one hour. This is hardly the case, because it assumes a new battery and perfect conditions. It also means the pilot was watching the ammeter at the time of failure. If you unwittingly proceed for a half hour after the failure, drawing energy solely from the battery, less than a half hour remains before power is depleted. As a rule of thumb, expect only fifteen or twenty minutes of emergency power.

Even that time span is misleading. Before the battery dies, it disables sensitive equipment as voltage drops. Under normal conditions, an alternator or generator maintains about 14.5 volts through the system. When the battery falls to about 11 volts, lights glow feebly and radios behave erratically. A press on a mike button no longer brings the reassuring click of send/receive relays, and communications are lost. Before these despairing symptoms appear, the pilot has two options that will enable him to survive nearly any electrical problem:

Detect the failure as early as possible.

When the battery silently takes over the electrical system it registers a warning on the ammeter. Don't miss it. Scan the instruments frequently at night.

Turn off every possible electrical item.

It is folly to run two identical navcom radios during electrical failure, because it seriously curtails remaining battery life. If you can't make an early landing, hoard the battery's amperes by flying by magnetic compass and flashlight. You want to conserve enough battery power to transmit an emergency call or obtain permission to land. To check navigation, turn on a VOR radio long enough to verify course or position, then turn it off.

Compass and flashlight notwithstanding, it's easy to get lost at night when all radios are shut down. If you want to leave one radio on, consider these options: If the plane is equipped with tube-type navcoms and ADF, the least drain is most likely from the ADF. It is a receiver only and draws little power. If navcoms are solid-state (they go on with no warm-up), they draw little power when supplying a VOR display. Tube-type navcoms need about 7 or 8 amperes, while the transistorized version reduces that current to 2 or 3. Thus the choice between a solid-state navcom set and a tube-type ADF is about equal.

How much current each radio draws is stated in the manufacturer's literature, or it can be estimated on the ammeter in a ground test. In an aircraft with a zero-center ammeter, turn on the master but don't run the engine. Turn each radio on and off and note how far the needle moves into the discharge region. This shows the relative number of amperes drawn by the radio. In aircraft with the other type ammeter (which shows total load drawn), operate the engine and switch each radio on and off. The amount of needle rise for each item tells the relative draw. From these readings you can judge which equipment to choose during power failure.

Dead battery in flight

A hapless pilot is one who discovers a power failure *after* the battery is dead. Now the situation demands quick disabling of every electrical item not needed to sustain flight. Removing the load might elevate battery voltage enough to drive one radio for brief navigational checks.

Even with a dead battery, all is not lost. Batteries of almost any type have remarkable resiliency if given a rest period. The principle is demonstrated when you forget to turn off a flashlight. If allowed to recuperate for several hours, it will light for a while longer. An airplane battery has similar regenerative powers.

This is what rescued the author in that electrical failure mentioned earlier. Every item was turned off until the plane arrived over the airport about thirty minutes later. It was urgent to contact the tower for landing permission because airline traffic was strung out along the final approach to one runway, as departing aircraft rose from another. (Fifteen minutes of circling for a light signal failed to attract the tower's attention.) The half-hour rest period revived the exhausted battery enough for a brief, but successful, radio transmission. Our plane received immediate clearance to land.

Engine failure

10

Shortly after a night take-off, says an old aviation chestnut, the motor of a single-engine plane runs rough. Fly over water and the airframe starts "oil-canning." Beyond gliding distance of land the engine will almost certainly throw a rod.

These hallucinations ride with a pilot on his first night solo and may not fade a hundred hours later. They are spawned by the greatest psychological deterrent to night flying: fear of engine failure.

Statistics on engines which stop for mechanical reasons provide some reassurance. Your chances of having a fatal engine-failure accident for reasons beyond your control are about one every 400,000 hours. In studying one year of 213 fatal night accidents, the NTSB cited the power plant as the cause in eight cases. That figure is inflated because some failures were probably due to improper maintenance. Flying a poorly maintained single-engine aircraft at night naturally balloons the risk. In the eight fatal accidents, these reasons were given for outright engine failure (followed by number of occurrences): magnetos (1); fuel system, i.e. vents, drains, tank caps (2); lubricating system (1); propeller (1); and undetermined reasons (3). These accounted for 3.76 per cent of all nighttime fatal accidents. In the same year the pilot was cited as the cause in over

90 per cent of all types of fatal accidents after dark. Raw statistics may not remove all the tension that flies with single-engine pilots, but they show that engines are extremely reliable—if not mishandled.

Flying a twin-engine airplane understandably cuts in half the risk of having an engine-failure accident. Yet NTSB figures draw a gloomy picture when a multiengine accident ends in a fatality. In studying the period 1965–69, the Board disclosed that chances of a fatal engine-failure accident in a twin were almost *four* times greater than in a single-engine plane. Losing an engine in the twin is far more likely to result in a stall or spin with deadly consequences. Poor pilot proficiency while managing the flight on the remaining engine seems to be the aggravating factor.

Fear of engine failure in a single-engine plane is not difficult to comprehend. Descending helplessly in the dark hardly fits a pilot's confident self-image. In the day he can plot a clever escape through some aerial back door when the weather deteriorates, or divert for a precautionary landing if oil pressure drops. But when the engine stops at night, these bits of flight insurance are canceled. There's no chance to slip steeply into a tiny clearing or bounce to a halt in a farmer's field. Engine failure at night is an aimless fall toward disaster. It doesn't have to be.

The problem of what to do in the five minutes or so after the engine stops to the instant the wheels touch, as we'll see, has been the target of considerable study. There are early-warning signs, restarting procedures, and other techniques to lessen serious injury in a crash. A remarkable program developed by the University of Illinois can even illuminate an emergency landing area with parachute flares, as shown in the illustration at the beginning of the chapter. In no case must a night pilot ride his plane toward a catastrophic crash without trying almost a dozen lifesaving tactics.

Rough engine

The first sign of impending trouble often appears on the gauges. Any significant change in pressures or temperature is reason enough for an early landing. A rough-sounding engine may also warn that failure is close at hand. Unless the manufacturer's manual gives other recommendations, try these steps to smooth a rough engine: Apply carburetor heat, check the mixture control for proper setting, switch fuel tanks or turn on a boost pump. Sometimes a faulty magneto disrupts ignition. Try to isolate it by switching to left and right mags to determine which setting produces smoothest engine firing. Spark-plug fouling is sometimes corrected by leaning the mixture for several minutes. As a problem emerges, tell someone on the ground your condition and position, then consider landing at the nearest lighted airport. To nurse a faltering engine, use minimum power to sustain flight.

There have been incidents of overcoming fuel blockage by continuous pumping of the primer knob. One pilot maintained 1,500 rpm in a light trainer and limped home to a successful landing. Try it if all other engine-restarting attempts fail, but be aware that it takes vigorous pumping to manually ram fuel into the cylinders.

Assume the engine resists every attempt at restarting. If there's time, transmit "Mayday" at least three times into the mike and announce your predicament and position. If your radio is not set to 121.5 MHz, the emergency channel, and you are too busy or anxious to turn to it, broadcast on the channel to which you are tuned. Chances are you'll be heard by someone.

Some pilots may not want to pay attention to the radio at a time like this, but it may be urgent to get a message on the air. A tragedy which haunts search-and-rescue officials is the large number of people who survive air crashes in light planes but succumb a day or two later because they couldn't be found quickly enough. If conditions permit, send out a call while you have good transmitting range at altitude.

The descent

As you call for help, trim the airplane for *maximum glide*. This speed, given in the aircraft manual, is usually quoted as an indicated airspeed with the propeller windmilling, flaps up and zero wind. Consult your manual for the appropriate speed, but here are typical figures given by manufacturers: Cessna-150 (65 mph); Cessna-172 (80 mph); Cessna-177 (80 mph); and Piper Cherokee-140 (83 mph). During the transition from cruise to glide, there may be a small height gain as speed is traded for altitude. In a few moments the airplane is descending at a rate of about 600 to 700 feet per minute.

What to do next is not official dictum, nor is it set down by aviation authority. Few pilots have ever faced the problem of night engine failure and there are few case histories. To fill the void, a synthesis of opinion was gathered from professional pilots ranging from FAA examiners and airline captains to instructors and charter pilots. An old-time flight instructor starts the speculation with a technique that's won a durable place in flying folklore. He first heard it from his instructor while flying one night over a dark area of New Jersey. "What would you do now," he asked the instructor, "if the engine failed?" The reply:

"That's easy . . . Take your right palm and press it firmly against your left palm, like this. Extend the fingers of both hands. Now repeat after me: Our Father Who Art in Heaven . . ."

Meanwhile, back on earth, other pilots would hold that in reserve until they tried more mundane techniques. A young, highly experienced instructor said he would glide toward the darkest area because it is probably the flattest terrain. There are no large obstructions, he reasons, to jut up and cast reflected light. Obstructions would stand out in shadowy relief against lighted portions of the terrain. Another flight instructor suggests avoiding any brightly illuminated area because it means population and obstructions. Given a choice, he would let down into a dark area but one that adjoins a lighted region to improve his chances of being seen and rescued.

A water landing, despite the flat, unobstructed expanse, is notably uninviting to professional pilots. Height over water is difficult to judge, even during the day, and there's the added hazard of ditching, then escaping from the cabin. One pilot suggests a water landing just beyond the surf line within swimming distance of shore. A fixed-gear plane, however, is apt to bury its nose wheel on touchdown and flip forward, submerging the cabin. Pilots have escaped this inverted position, but the added stress of darkness could prove overwhelming.

Landing on a highway at night sparks lively controversy. A road simulates a lighted runway, but it's crossed by overpasses and lined with wires. One FAA examiner said he would not attempt a highway landing unless he was certain of wingtip clearance. He never had an actual experience, but he suspects that an aircraft with a 30-foot wingspan would not fit on a two-lane highway, though it might land on a wider road.

An airline pilot believes a highway landing might prove successful. Interstate highways, he points out, have access ramps and areas that might accommodate an emergency landing. Touching down on the main section of an interstate highway, he said, requires a special technique. He would bring the plane down at higher-than-normal approach speed and position it behind what he calls a "leader" car traveling about 60 mph. During this time, cars following the "leader" would see the plane and its flashing anti-collision lights in time to slow down. As the plane moved within about two car lengths of the leader, the pilot would dissipate excess flying speed and let down to the ground.

Touchdown

In a special study for the NTSB, Gerard M. Bruggink made several eye-opening statements about emergency landings. Bruggink, an accident investigator, disagreed with what he called the "school solution" to emergency landings. In the conventional method, students are taught to place too much stress on finding a "suitable landing" area to save the *aircraft*. This conditions a pilot to foolishly press on in deteriorating weather because he sees no "safe sites" for a precautionary landing. In a considerable number of fatal accidents a desperate pilot loses control in IFR conditions.

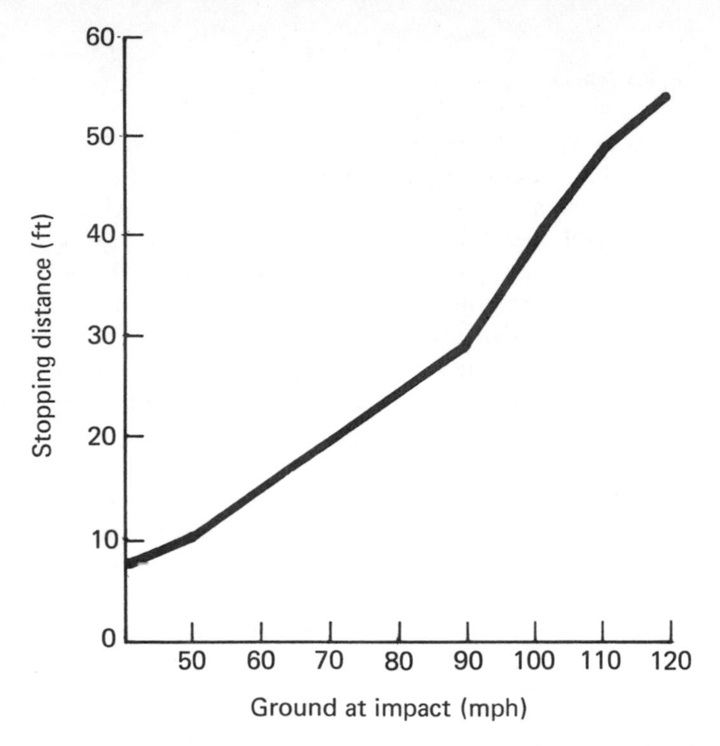

92. Stopping distance vs. groundspeed during *uniform* 9G deceleration.

Bruggink argues against the "suitable landing area" in declaring that "almost any terrain can be considered suitable for a survivable crash landing if the pilot knows how to use the aircraft structure to protect himself and his passengers." Get the airplane on the ground—almost any ground—and chances are good that the pilot will survive.

His study is also a veritable bible of what to do just before touchdown at night with a dead engine. The most important technique is simply:

Land at the lowest possible, but controllable, airspeed.

Out of a mass of technical data this statement sums up the greatest lifesaving technique in a forced night touchdown. The mathematics of high landing speed are horrifying. If ground speed *doubles,* the destructive force in a crash zooms by *four* times. To prove the value of landing against the wind, the report gives this example: If touchdown is made at 60 mph, what speed does it take to *double* the damaging force? The answer a mere 85 mph.

As touchdown occurs, the next critical factor is getting rid of G-forces over the shortest ground run. This gives the plane less time to strike obstacles at great impact. Again, speed is the killer. A typical light plane is designed to protect occupants in crash landings which generate up to 9Gs in the forward direction. Using that as a reference, you can see in Figure 92 what happens with small speed differences. For example, if the plane touches down at 50 mph, 9Gs can be consumed in less than ten feet! At 100 mph, it takes nearly forty feet to accomplish the same effect.

High-speed forward motion isn't the only threat. At the instant of touchdown, vertical speed drops abruptly to zero. This quick change in velocity is a common cause of spinal injury. According to the NTSB report, it results from the most critical and inexcusable error in any emergency landing—the pilot's loss of initiative over the aircraft's attitude and sink rate at touchdown. Not only does the human body have poor resistance against vertical compression, but the aircraft doesn't offer much protection in this direction. If the plane descends at a rapid rate, say more than 500 feet per minute, and contacts earth in a flat attitude, little structural deformation cushions the impact. Shock is transmitted directly to the passengers. (This is especially true in gear-up landings because there is no undercarriage to give way.) After a fast vertical descent in a high-wing plane the overhead structure may collapse on the passengers. Where terrain is soft, a rapidly descending aircraft is more likely to dig its nose into the turf and convert a rapid descent rate into crippling deceleration that pitches the passengers forward. A return to earth at the lowest forward and vertical speeds—short of a stall—slashes these forces.

Despite the obvious value of the NTSB report to night fliers, it made no specific references to emergency landings in the dark. After its publication Bruggink introduced fresh insights into the problem. He remarked that a landing light "could make all the difference." The last fifty feet before touchdown he considers the most critical for avoiding the worst possibility—a head-on collision with a solid object. If the landing light picks out an obstacle, limited, evasive maneuvering may be possible. *Limited* because steep banks increase stalling speed, and controlled flight is the supreme goal.

Bruggink took exception to a widely circulated suggestion that an airplane gliding toward an emergency touchdown in a confined area, where a reasonable landing is impossible, should fly toward a space between two trees. The theory is that impact danger is reduced as the wings shear off, leaving cabin and occupants intact. As an accident researcher, Bruggink agrees with the idea of sacrificing the plane in favor of passengers, but not in the air. It's too dangerous and uncertain. "If heavy tree-trunk contact is unavoidable once the aircraft is *on the ground,*" he says, "it is best to involve both wings simultaneously by directing the aircraft between two properly spaced trees." He emphasizes "on the ground."

In the moments before touchdown other techniques reduce the danger. There's less chance of fire by pulling the mixture control to the cut-off position and turning the fuel valve to an off position, if there is one. Turn off all switches except the master; you want the landing light on. Crack open cabin doors, because the impact might distort the airframe and jam the door. Place an object, possibly a shoe, in the door opening. Have your passengers assume the crouching, head-holding position recommended by airlines to reduce vertical impact on the spine and to cushion the head against forward impact. Don't lose flying speed or control until the last possible moment.

Flare-assisted landing

On an April evening in 1955, a small airplane droned through a mid-western sky. Suddenly, without a warning cough, the engine died. The pilot trimmed for an emergency landing in the dark, but as the plane glided to 2,000 feet the sky exploded with a brilliant cone of light. The plane dove and curved into the glowing apex above a widening circle of light on the terrain. The pilot could see the ground for a mile in every direction. He sighted an emergency field, landed and walked away unharmed. The plane was unscathed.

This was one of fourteen experiments conducted by the University of Illinois to discover if flares could make night flying less threatening to the single-engine pilot. Military services had used parachute flares to light up drop zones at night, and similar techniques might apply to civil aviation. The University believed more use would be made of light aircraft at night if standard procedures could be developed. Since there was little available data on flare-assisted landings, the University devised its own study and started flight tests in 1955.

The results were remarkable. Every one of thirteen participating pilots believed the technique would enable him to survive a forced landing at night. Out of fourteen trials, thirteen were successful—the sole failure caused by human, rather than equipment, error. One pilot had arbitrarily flown a right-hand approach pattern when a left-hand pattern had been agreed upon in advance. It caused him to overshoot the field. This hardly marred the overall result. From these experiments emerged a proven weapon against nighttime engine failure.

The study also revealed certain disadvantages. Flares are costly and an extra item to carry aboard the airplane. They must be returned every few years to the manufacturer for inspection to insure reliability. And much of the technique's success is tied to careful procedures, practice in the air and pilot proficiency. Since the methods worked out by the University provide a rare, positive solution to nighttime engine failure, they're worth considering in some detail.

The problem was outlined in the form of three questions: How much area and detail can be seen from a flare's illumination? What are the best launching altitudes? And how would a pilot maneuver to get the most benefit from a flare's light? Forty parachute flares were fired in simulated engine failures high over the University's airport. Ground lights were extinguished and pilots could rely only on flares and aircraft landing lights for guidance. The pilots selected for the test were mostly flight instructors, and only one had had previous experience with flares. One had experienced power failure (a partial one) at night that led to a forced landing. The airplanes were an assortment of small, single-engine Beech, Piper, Cessna and Stinson craft.

The flares were of several types. Nearly half were Class 3, a 1-minute pistol-fired type (Fig. 93) that emits white light; a dozen were Class 2, 1½-minute electrically fired flares (Fig. 94); and a dozen were 30-second red signal lights (37 mm). When launched from an aircraft, the flares would fly out horizontally about 75 feet, then descend no faster than 550 feet per minute on a parachute of 4 to 5 feet in diameter. Light intensity ranged from 70,000 to 110,000 candlepower. A flare would be expected to extinguish before reaching the ground if released no lower than 800 feet above ground for the 1-minute model, or 1,000 feet for the 1½-minute flare.

As each test plane arrived over the darkened airport, an observer pulled off the power and announced "forced landing" to the pilot. From that moment the pilot would be allowed to fire two flares. The first would illuminate a general area to help him maneuver into a favorable position for an approach. The second flare would reveal further detail needed to make the landing.

Early in the tests it was clear that a pilot stricken with engine failure at night doesn't need to light up the sky from the instant the prop stops to touchdown. In seven trials landings were successful—in fact, without firing the second flare. The first flare enabled the pilot to pick a field and maneuver into position for the landing. In the course of the tests researchers remarked: *"The landing lights will give enough illumination to complete the landing with a reasonable chance of avoiding major obstacles on the ground."* (Italics mine) This finding is especially important because it offers night fliers equipped *only* with a landing light a rudimentary technique for avoiding a direct strike. As stressed in the NTSB report mentioned earlier, a head-on collision is the worst possibility.

The University pilots also found that flares rated to drop 550 feet per minute actually fell more slowly and allowed the plane to descend for a better view of the ground. A pilot could quickly slip under the cone of light to win a longer period of illumination.

Wind drift occasionally proved tricky. A 15-mph wind carries a 1-minute flare one-quarter mile downwind from the drop point. Aware of this movement, the pilot always reaches for a field in a generally downwind direction. Drifting becomes more critical after the second flare is launched in the final phase of landing. Unless it is released upwind of the site, the flare may float beyond the landing area while the plane turns on final approach. Pilots solved its worst effects by planning a 180-degree or 360-degree overhead approach to keep the second flare over the field during the touchdown.

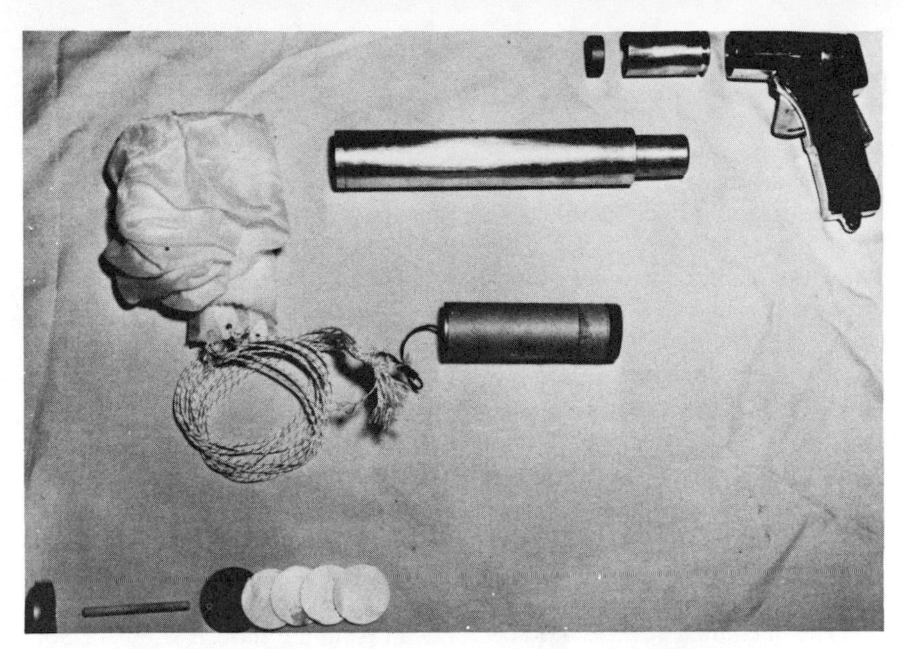

93. Class 3, 1-minute pistol flare. Parachute is at left, pistol at right.

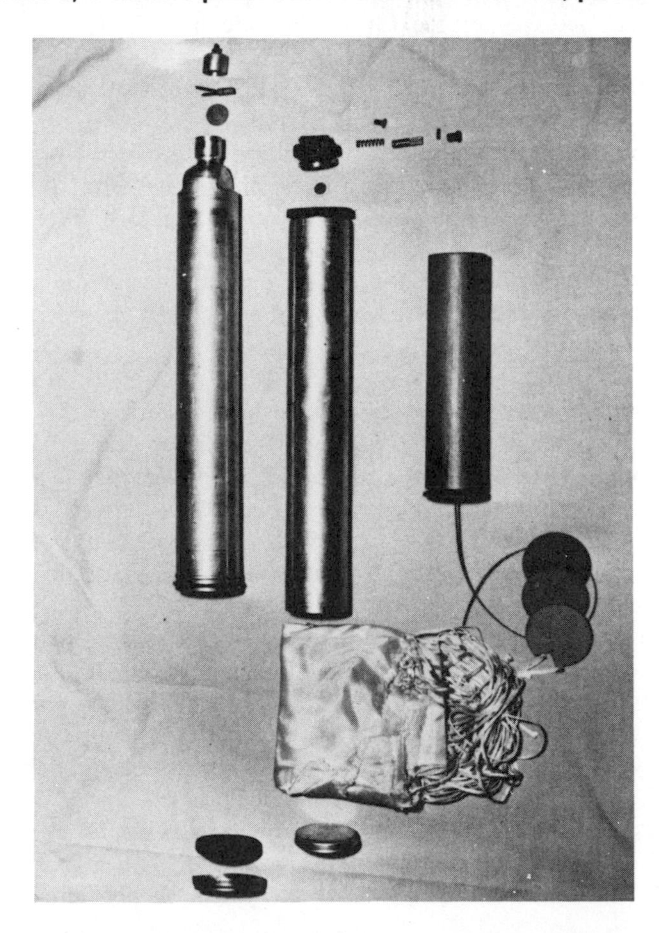

94. Class 2, 1½-minute flare shown in its separate components. Parachute is at lower right.

The first three trials by the researchers failed. They assumed that a flare should be launched from an altitude which yields greatest burning time, causing the light to die as the flare touched the ground. Practical flights proved otherwise. As a flare descends, its two important quantities—area of illumination and brightness—vie in a conflict controlled by altitude. The higher the flare, the greater the region it illuminates. As it nears earth, the cone grows smaller but more intense. During early runs, second flares were dropped too low and out of range to be of much aid in the landing. Further trials suggested that the second flare should expire some 400–500 feet *above* ground when the plane touches ground. The higher flare altitude also widens the margin for drift error.

From their experiments, the researchers developed these basic guidelines:

Release flares upwind of the desired area to permit a 180-degree or 360-degree overhead approach;

The maximum release altitude for a 1½-minute flare is 2,000–2,300 feet above ground;

Maximum release altitude for a 1-minute flare is 1,200 feet above ground;

Minimum release altitude for a 1-minute flare is 1,000 feet above ground.

If these altitudes are observed, the pilot discerns light-colored objects (like concrete and white buildings) from 2,000 feet with a 1½-minute flare. The objects are visible at distances up to a mile in every direction from the flare. A smaller, 1-minute flare dropped at 1,500 feet shows the same objects but only to three fourths of a mile in every direction. After firing the second flare, just before landing, the pilot may see fences, ditches, small slopes and other details.

Flare types. Manufacturers supply flares which fasten to the aircraft or are carried as portable, pistol-fired units. The former type is launched by pressing a button (Fig. 95) in the cockpit. This convenience is partly offset by the need for an installation of racks (Fig. 96) and wiring by an aircraft mechanic. A departing flare kicks with a recoil of about 150 pounds, and the manufacturer's instructions must be carefully followed to avoid damage to the aircraft.

Technical problems during the University program were mostly due to aging flare powder or a slight deformation of projector tubes. Nearly all malfunctions from these causes are eliminated by renovating flares at least once every three years. In the University's experience, only one flare less than three years old failed to fire. Moisture absorption probably caused it.

The pistol flare (Fig. 97) has the great advantage of portability. It can be carried aboard different aircraft by renters, club members, and other pilots who don't always fly the same plane. A problem discovered

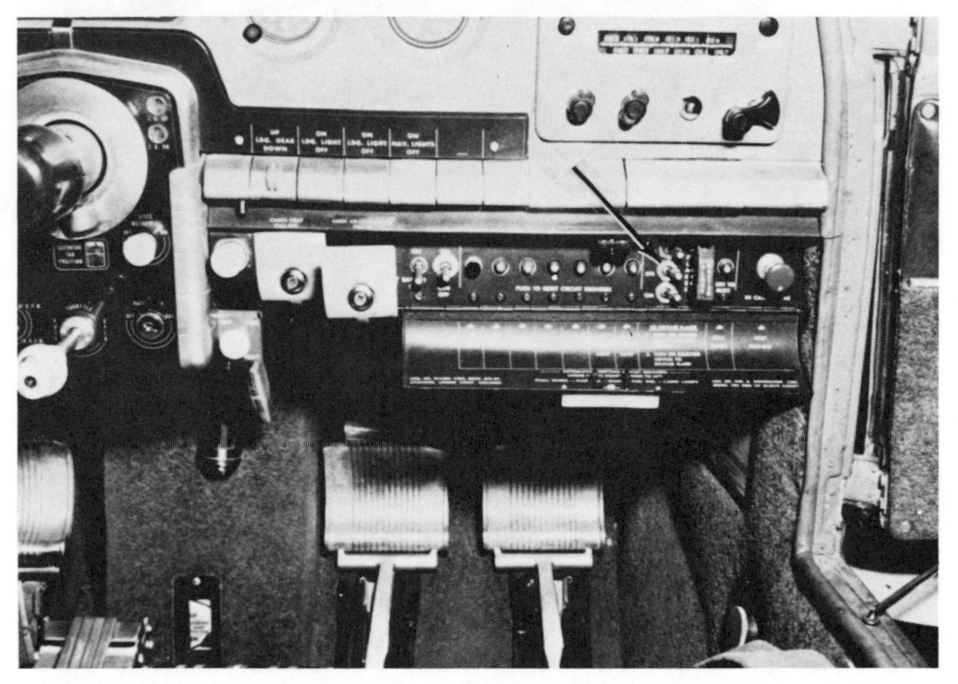

95. Flare-firing switches are on lower right of instrument panel (see arrow).

96. Rack-mounted flares in baggage compartment of Beechcraft Bonanza.

97. Loaded pistol flare is held through open window using straight wrist and elbow position. Left-hand grasp allows shoulder to absorb recoil.

by researchers was the flare's accessibility in the cabin. When the flare container was placed on a shelf in the baggage compartment, the pilot averaged twenty-four seconds to retrieve it, load the pistol and fire the flare. The practical solution is to reduce the size of the carrying case (which normally has five flares) to carry only two flares and the pistol (Fig. 98). After take-off the case is placed on the floor between the pilot's legs. A lanyard from the case is placed around the control yoke and the pistol attached onto one of the flares. During an emergency the pilot can grip the case between his knees, use one hand to fly the plane, and the other hand to hold the pistol.

Most pilots are right-handed, which proved awkward for pistol-firing in flight. Since the pistol is pushed through an opening at the left, a right-hander can't align elbow and wrist to absorb the recoil. With the left hand, however, a straight wrist and elbow transmit the recoil to the shoulder (Fig. 99), where it's easily absorbed. Most pilots barely noticed any recoil in the tests.

The art of firing a flare depends on the specific airplane. First consideration is, of course, not to strike the aircraft. Secondly, a flare surrenders most light if it leaves the plane as horizontally as possible. In some planes a clear shot calls for a forward and downward trajectory; in others, a backward and downward line of fire satisfies the launch requirements. The small window of some Pipers, for example, afforded too little area for the pilot to aim the flare along a clear path. This was solved by removing a hinge pin at the window's base and replacing it with a removable pin (Fig. 100). It allowed the window to be removed

98. Pistol-fired flares in portable case. Pistol is at left, two flares at right.

99. Firing pistol flare through open window of Piper Tri-Pacer. Left hand is used, with wrist and elbow held in straight line.

in flight for a much larger opening. Once the loaded pistol is thrust through the opening, the pilot slows the plane to a normal approach speed to prevent an unsteady aim in the wind blast.

The University team suffered no misfires in launching sixteen 1-minute pistol flares. It did predict possible trouble if a flare doesn't fit easily into the pistol barrel. Too much snugness is corrected by removing high spots on either flare or barrel with emery cloth. If a flare is forced into a barrel, an expended shell may jam and prevent the pilot from pulling it

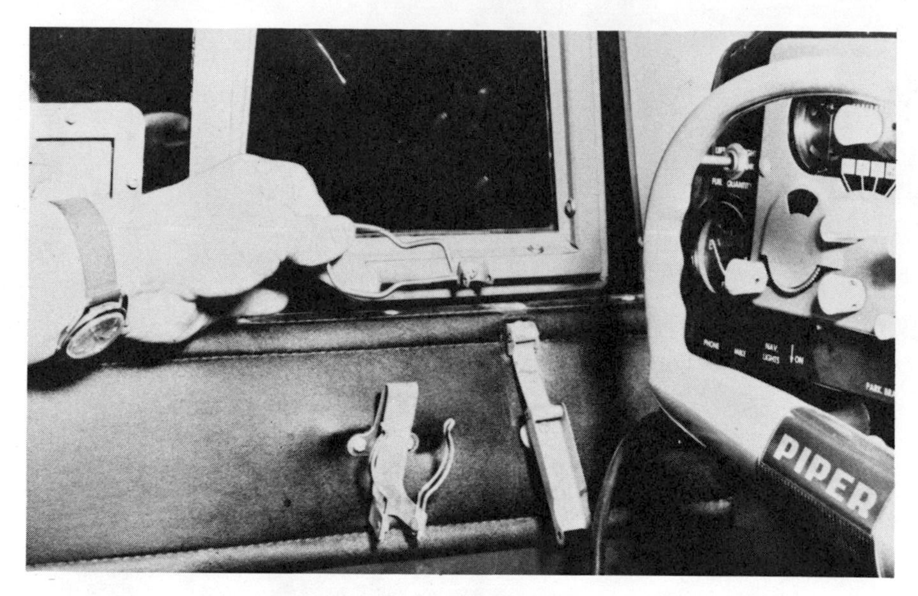

100. Window in Piper Tri-Pacer is modified with hinge pin to allow opening for flare pistol.

out. No hangfires, where the flare fails to ignite immediately, were encountered, but it is a potential hazard. To prevent fire inside the plane, any flare which fails should never be drawn back into the cabin. Some pistols have ejectors to get rid of an unfired flare.

There are several pistol and hand-launched flares offered to the private pilot which should never be used for landing. They are *signal* flares with rated intensities of between 12,000 and 20,000 candlepower. Although visible to the eye at tremendous distances, often more than twenty miles, they serve only to attract attention. Illumination is far too feeble to light up the terrain for landing. The pistol and other flares in this test program range from 70,000 to 110,000 candlepower.

In an attempt to communicate the total experience of a nighttime engine failure and how to avoid disaster, the experimenters created the following hypothetical case. You're flying a Beechcraft Bonanza at 4,000 feet over unfamiliar terrain when all of a sudden . . .

The engine stops. You try to restart by checking gas selectors, carburetor heat and magnetos. Descending now at 740 fpm with a dead engine, you're committed to a forced landing. This is Phase 1, a period that commences when the prop stops and ends when the plane is 2,000 feet above ground. It is useless to fire a flare now, because the light would prove too feeble at high altitude to illuminate the ground. In the descent from 4,000 to 2,000 feet above ground, you'll have about three minutes to

101. Flare illumination during Phase 2. First flare is shown drifting from left to right, as pilot descends below flare and circles to choose field.

complete Phase 1. Slow the airplane to normal approach speed with the trim tab. If you're not yet at 2,000 feet, study the ground before launching the first flare. The slight ambient light of even a dark night may reveal a general area. Densely populated regions expose themselves by city lights, and shadows may set off hills. Once the general area is chosen, Phase 1 should end, if possible, with the plane upwind of it. Head into the wind by the time the altimeter indicates you are 2,000 feet above ground.

Phase 2 is beginning. At 2,000 feet launch the first flare. Its intense light floods the countryside, giving you a chance to appraise possible landing areas. Since the flare rocketed from the airplane's left side, turn to the left and dive under the flare to avoid looking at its blinding light and to get the best view of the terrain. When you spot a likely area, try to maneuver upwind of the site.

Phase 2 is barely more than a minute's duration. It's limited by the flare's burning period and the altitude at which the second flare must be launched. That height is 1,200 feet above ground. During the brief interval it should be possible to make one 360-degree spiraling turn (Fig. 101) at double the standard rate, or a 180-degree turn with an additional straight run of about three fourths of a mile. Turns should be moderately steep—about 30 degrees of bank—to avoid losing time. No matter where you are at the 1,200-foot altitude, Phase 2 is ended.

102. Flare illumination during Phase 3. Second flare is seen drifting from left to right. Flare has been fired (left) extreme upwind from selected field. Airplane circles and touches down before flare extinguishes (right).

The second flare opens Phase 3. The period begins at 1,200 feet and ends when the airplane is on the ground (Fig. 102). Since you are headed into the wind, turn the airplane around and fly downwind in a maneuver not unlike a normal landing pattern. The entry into the pattern is similarly conventional. Approach the downwind leg at a 45-degree angle. Meanwhile, the second flare should be floating down, exposing a considerable amount of ground detail. If the airplane has retractable gear, decide now whether to make a wheels-up or wheels-down landing. (The University recommends wheels up if there's the slightest doubt about inadequate length or smoothness of the field.) Keep the pattern tight, turns steep, and estimate when and where to turn base and final. Turn on the landing lights on the base leg and decide how many degrees of flaps are necessary. (There was a tendency by the test pilots to level off too high under the light of a flare. They went to the other extreme when this was called to their attention and tended to touch down before rounding out.)

In the final moments of landing, the descending flare burns overhead at an altitude of about 500 feet. Now you can see, and possibly avoid, boulders, ditches and fences. If the flare expires during final approach, landing lights should pick up obstacles from about thirty feet aloft.

The success of a flare-assisted landing is reflected in remarks of the test pilots who participated in the program. Each felt he could now walk away from a nighttime forced landing. But, they add, it can't be easily done without practicing the maneuvers during the day and flying dry runs at night with the co-operation of the local airport. (A summary of each phase is shown in Table 1, with additional recommendations in Tables 2 and 3 at the end of this chapter.)

Following its study, the University of Illinois equipped its own aircraft for flare-assisted landings. The planes fly extensively at night for flight training. In nearly twenty years, however, the researchers failed to personally confirm their findings in thousands of hours of single-engine night flying. They never had an engine failure!

TABLE 1 Guidelines for flare-assisted landing

(Altitudes shown are for 1½-minute flare activated from instrument panel. Altitudes shown in parentheses are for pistol-fired flare.)

Phase 1

1. This phase begins with the actual engine stoppage and ends at 2,000 (1,800) feet above ground. If the engine stoppage occurs below 2,000 (1,800) feet, the entire phase is omitted.
2. Using night vision only, the most desirable general area is selected for the forced landing.
3. The airplane must arrive over the selected area not below 2,000 (1,800) feet. It is desirable, but not mandatory, to be headed into the wind.
4. Any additional time (altitude) should be devoted to preparing the airplane for the actual landing. This task will vary in different types of airplanes but should include such things as adjusting trim, cutting unnecessary switches, setting fuel-selector valves, and bringing the propeller to a horizontal position with the starter.
5. If the airplane has retractable gear, all such adjustments should be based on the assumption that a gear-up landing will be made.

Phase 2

1. This phase begins at 2,000 (1,800) feet and ends at 1,200 (1,000) feet above ground.
2. The first flare is fired at 2,000 (1,800) feet and is used to select a specific field within gliding distance (approximately 1½ miles) of the point of the flare release.
3. The airplane must arrive at the extreme upwind end of the selected field by the time it has descended to 1,200 (1,000) feet above ground.
4. The field-selection search should be made using a left turn and keeping the airplane within the flare's cone of illumination.
5. During the positioning process (maneuvering to the upwind extremity of the selected field) the pilot should mentally note the type of overhead approach which the 1,200 (1,000)-foot heading will necessitate.

6. In general, fields selected in the upwind portion of the cone of illumination will necessitate a 360-degree overhead approach; fields selected in the downwind sections of the cone of illumination will dictate a 180-degree overhead approach; and fields selected in the central portion of the illuminated areas will require combinations of the 180- and 360-degree overhead approaches.

Phase 3

1. This phase begins at 1,200 (1,000) feet and ends with the completion of the landing roll.
2. The second flare is fired at 1,200 (1,000) feet above ground and is used for the approach and landing. If the flare is fired above 1,200 (1,000) feet, it will very likely extinguish before the landing is accomplished. Firing altitudes below 1,200 (1,000) feet are likely to result in the flare's being out of usable position for the final approach.
3. The triangular-shaped overhead approach is almost mandatory, since the approach and landing flare must be fired at the upwind extremity of the selected field.
4. The all-important decision on gear position must be made during Phase 3 in sufficient time to get the gear down if the pilot elects that configuration.
5. Landing lights should be turned on before the flare extinguishes, probably on base leg. Flare extinguishment before touchdown does not necessarily mean a complete failure on the problem.
6. The third, or reserve, flare should be fired in case of misfire of either of the two planned flares.

TABLE 2 Recommendations based on fourteen simulated night-time forced landings using specific airplanes and specific flares

Flare Type	Class 2 1½-minute Electrical	Class 3, 1-minute Pistol Flare			
Airplane Type	Bonanza C-35	Piper Tri-pacer	Cessna 170	Cessna 140	Stinson 150
Power setting used	Idling	Idling	Idling	Idling	Idling
Indicated airspeed mph	95	80	80	70	85
Passenger load including pilot	4	4	4	2	4
Landing gear	Lower on base or final as necessary	Down	Down	Down	Down
Flaps	Lower on base or final as necessary	Lower on base or final as necessary	Lower on base or final as necessary	Lower on base or final as necessary	Lower on base or final as necessary
Airplane's average rate of descent in feet per minute	740	870	670	640	860
Altitude above ground to release flare for field selection purposes	2000	1800	1600	1500	1800
Average time from release of first flare to altitude for second flare	67 seconds	55 seconds	57 seconds	59 seconds	55 seconds
Position of airplane relative to selected field for release of second flare	Extreme upwind	Extreme upwind	Extreme upwind	Extreme upwind	Extreme upwind
Recommended approach pattern from second flare to landing. Overhead approach.	180 or 360 degree	180 or 360 degree	180 or 360 degree	180 or 360 degree	180 or 360 degree
Desirable airplane position in relation to flare to obtain best visibility	Below	Below	Below	Below	Below
Time from second flare to extinguishment	1½ min.	1 min.	1 min.	1 min.	1 min.
Excess altitude necessary to dive off immediately following release of second flare to land before flare extinguished	200 ft.	100 ft.	400 ft.	400 ft.	150 ft.

TABLE 3 Summary of simulated nighttime forced landings developed during trials at the University of Illinois

Airplane	Flare Type & No.	Moon Phase	Official Weather		Unexpected Experiences	Landing Completed by Flare or Landing Lite	Total Time Flare #1 to Landing	Landing Successful
			Ceiling and Visibility	Wind				
Bonanza	1½' 17875 17874	Dark	Clear 15+	SSW12	Defective flare master switch	Landing lite flare #2 drifted out of position	2:58	Yes
Bonanza	1½' 17420 17421	¾	E250 ⦷ 15	NNW8	Flare #2 malfunctioned, no illumination	Landing lite	2:32	Yes
Bonanza	1½' 17876 17422	¾	E250 ⦷ 15	NNW10	#2 flare released too late	Landing lite flare #2 drifted out of position	3:03	Yes
Cessna 140	1' 6804 6805	Dark	30 ⦶ E50 ⊕ 10	SSE10	Did not have gun loaded with #1 flare	Flare	2:21	Yes
Cessna 140	1' 6802 6803	Dark	35 ⦶ E50 ⊕ 10	SE11	#1 cartridge hard to remove	Flare	2:20	Yes
Cessna 170	1' 8336 8345	Dark	10 ⦶ E50 ⦷ 8	N10	Right hand pattern, overshot landing	No landing	No landing	No
Cessna 170	1' 6807 6806	Dark	10 ⦶ E50 ⦷ 6HR–	N10	Fumbled unloading and reloading pistol	Flare	2:16	Yes
Piper Tri-pacer	1' 8334 8337	Dark	80 ⦶ 12	E10	Flare hypnosis. Failed to reload pistol at proper time	Flare	2:05	Yes
Bonanza	1½' 19003 19002	Dark	80 ⦶ 12	E10	#2 flare failed to leave plane	Landing lite	2:28	Yes
Bonanza	1½' 18923 18924	Dark	8 ⦶ M22 ⊕ 15R–	W15	None	Flare	2:54	Yes
Stinson 150	1' 8341 8335	Dark	60 ⦶ 12	NW8	Difficulty removing first cartridge	Flare	2:16	Yes
Cessna 170	1' 8338 8340	Dark	E100 ⦷ 15+	SE7	Fumbled reload. Flare released wrong altitude and position	Landing lite flare #2 extinguished on final	3:00	Yes
Cessna 170	1' 8342 8344	Full	E80 ⊕ 10	S6	Flare extinguished just before landing	Landing lite flare #2 extinguished on final	2:47	Yes
Bonanza	1½' 19001 18925	Full	E80 ⊕ 10	S6	#2 flare failed to ignite	Landing lite	2:59	Yes

Log of a night flight

11

The FAA's monotonous cant about preflight planning becomes a night pilot's formula for survival. The cabin of a small plane, strangely insulated from the earth by a bubble of blackness, is a hostile place for impromptu navigation. The flight is vastly safer, less hectic and more enjoyable if it is first flown on the ground. With every detail neatly logged, ready to unfold on your lap, a long cross-country in the dark is an exhilarating experience.

Let's fly an actual trip. It will begin moments after the sun's disk touches the horizon, continue into a deepening twilight, and end near the stroke of midnight. It starts at Danbury Municipal Airport located in Connecticut some fifty miles northeast of Manhattan (see map in Fig. 104) and traces a complex zigzagging course to seek out a diversity of nighttime conditions. Soon after take-off, the plane will head northwest toward jagged terrain where Catskill and Berkshire mountains begin a sloping rise north of New York City. The flight then turns eastward over an unlighted, almost unpopulated region before an approach and landing into an international jetport. We will attempt to home on a weak radio beacon en route to Long Island Sound, a body of water which cleaves the land mass between New York and New England, and cross that black

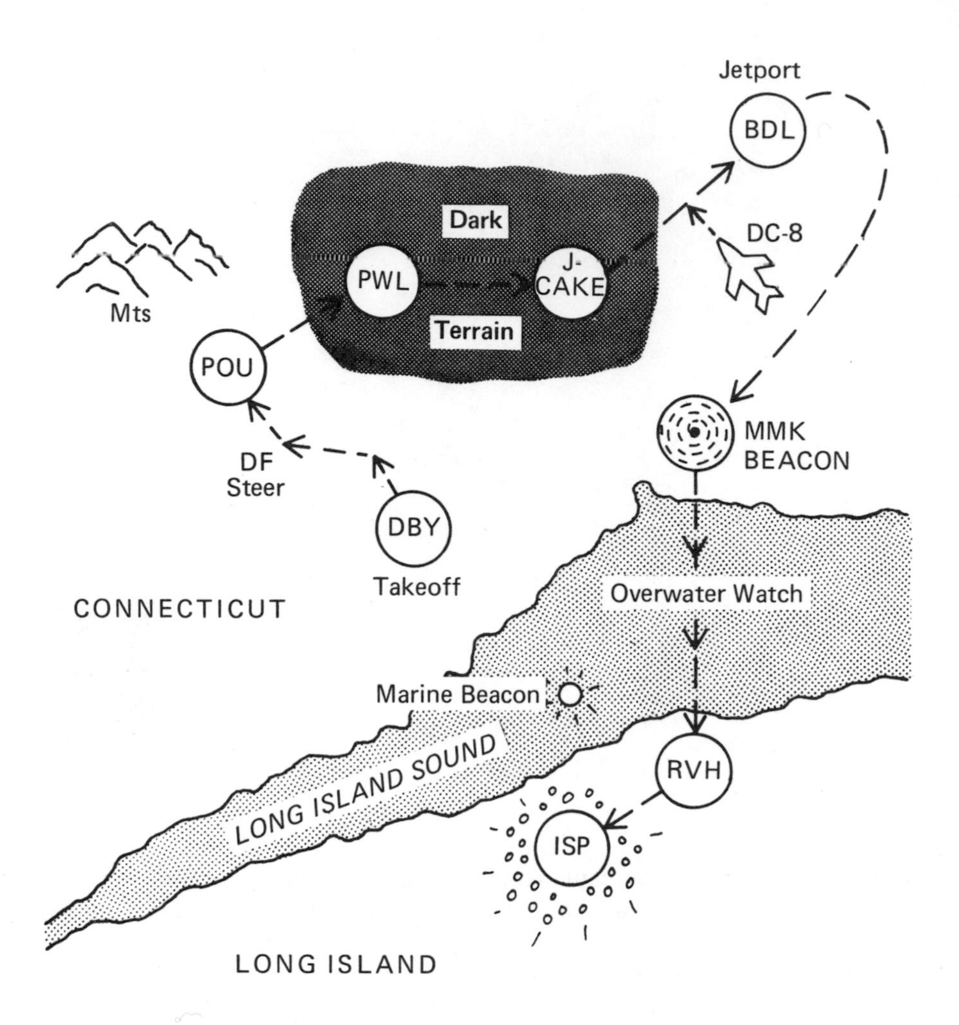

104. Route of the night flight.

void at its widest point. At 11:00 P.M. our plane will try to pluck out a regional airport embedded in a constellation of interstate highways and congested suburbs. With two souls aboard, our single-engine Cherokee will find its way by radio, stopwatch, compass, eyeball and, maybe, by sighting the stars.

The trip begins twelve hours before take-off. At 8:00 A.M. a small portable radio at home tuned to the National Weather Service brings in the earliest picture of weather over the region.

"At 5:00 A.M.," announces a taped voice, "the weather map shows a ridge of high pressure extending from Ontario to central New York that is expected to move eastward off the mid-Atlantic coast by Wednesday (the next day) bringing increasing cloudiness. Fair tonight, with 10 per cent chance of rain today and tonight, 20 per cent on Wednesday."

A reasonable outlook for VFR night flight. No frontal movements or violent weather lie within two hundred miles of the route, and the only concern is the ceiling. To maintain good clearance over mountainous terrain and fly the overwater leg, we want no clouds lower than 5,000 feet. It is still too early to predict the ceiling with certainty.

Over morning coffee, a glance at the weather map in the newspaper delivers a broader picture of weather systems over the country. There is only superficial interest in this chart, because it was prepared the night before and is too vague to forecast VFR conditions for a light aircraft over a short haul. But the newspaper carries other valuable details affecting the flight. Most important, the sun will set at 8:18 P.M. Since winds are expected to blow from the west and our airport has a westerly runway, a take-off at sunset would place the glaring disk directly in the windshield. We can cope with this by holding two fingers in front of the eyes to mask out the sun, a technique that's better than shielding with a whole hand, because it obliterates less of the runway. Blinding sun in the eyes would probably not be a hazard on this flight because take-off time is scheduled for a half hour after sunset. This would expose us, however, to a peculiar quality of twilight that often disturbs a novice night flier.

The morning newspaper also reports movements of heavenly bodies. That night, three bright travelers will cross the early evening sky—Mars, Jupiter and Saturn—all visible in the west until midnight. If man-made instruments should fail, we might steer by the celestial system known before the Middle Ages! A half circle pictured in the paper predicts that the moon will be in its first quarter and, if no clouds block the lunar glow, possibly lighten the terrain.

By early afternoon a radio in the office pulls in the transcribed aviation forecast from a radio beacon sixty miles away at Newark International Airport: "A weak high-pressure ridge will move from eastern Ontario southward over Ohio and eastward over central New York by late evening." High pressure promises good weather, but a new word—"weak"—suggests that cloudiness might modify our route. But there remains enough encouragement to start the flight log. For the first leg, Danbury to Poughkeepsie, the log will contain four items of crucial interest to any VFR night pilot: magnetic course (MC); distance of the leg in statute miles (sm); estimated time en route, in minutes (ETE); and height of obstructions above sea level (MSL). Here's how the first entry in the log appeared:

DBY–POU

MC—335°
Dist—27 sm
ETE—13″
Obstructions—1,500′

The magnetic course (335°) is extremely important because the compass is most resistant to failure. Couple it to ETE (13 minutes) and, if radios expire, we can navigate to the vicinity of the first airport (27 sm, or statute miles, away) and look for the rotating beacon. The last entry, Obstructions, means the sectional chart shows something projecting 1,500 feet (MSL, as indicated on the altimeter) into the flight path within about ten miles on either side of the course. This entry is significant because most obstructions simply never appear at night. The number also limits how low you can safely fly without striking an obstacle, if clouds force you to lower altitudes. Obstruction figures are difficult to discern on a chart in a darkened cabin while in flight.

Once these basics are logged, you may choose to fly every leg with the greater ease and precision of VOR or ADF radionavigation. With the dead-reckoning calculations already done, the log stands by as a reassuring backup. What if the *backup* fails? Carefully plotted courses and leg times, too, can disintegrate under stressful conditions—a low-time pilot on a very dark night, or losing your paperwork in the Stygian depths under the seats, for example. To stimulate this predicament, our Cherokee will take off on its first leg and temporarily abandon the security of flight log, stopwatch, compass, and navigation receivers. We will raise a Flight Service Station on the radio to ask to be guided by a "DF steer."

One hour before flight time a phone call to weather verifies that a stalled low-pressure area in the Midwest poses no threat to VFR conditions this evening. The sequence reports are a bit less encouraging. At 8 P.M. they show moderate cloudiness starting to form over the major terminals along our route. Poughkeepsie, point of the first landing, reads 5,000 scattered, 15 miles visibility; Bradley International, 5,500 scattered, and 15 miles; Islip, Long Island, 5,000 scattered, and 15 miles. Islip bears watching. At our time of arrival the Islip ceiling is forecast to go to 5,000 broken, and cloud heights below that figure would endanger a prudent crossing over open water in a single-engine craft.

The Cherokee moves quickly over Runway 26 at Danbury and ascends into the half-light of a spring evening at 8:45 P.M. The take-off is extraordinarily easy. The solar disk is already behind a hill, but the runway surface remains clearly visible in the dimming atmosphere. The Lycoming engine pulls in drafts of dense night air and roars with satisfaction.

Take-off into the twilight sky can be unsettling to the first-time night pilot because the earth's swing from day to night generates an alarming illusion that visibility is dropping to zero. It's an optical trick apparently brewed from a curious mixture of light at dusk. After the sun drops below the horizon, it scatters blue light high in the atmosphere. At the same time, it also projects horizontal rays through hundreds of miles of dust to produce the characteristic red glow of sunset. As red and blue light rays mix, a purple gloom infiltrates the countryside like sea fog rolling over a coastal village. The illusion intensifies as cooling air precipitates trails of haze between the hills. But the scary prospect of zero visibility usually expires with the dying sun. The advancing night is animated with pinpoints of light that pierce the darkness from fifty or sixty miles away to prove that visibility may, indeed, be excellent.

Our Cherokee is now at 4,500 feet heading in a northwesterly direction, catching a final glimpse of daylight mirrored on the lakes below.

"Poughkeepsie Radio, this is Cherokee 6654J requesting a practice DF steer."

The call draws a cheerful response from a Flight Service Specialist seated in a small building snuggled in the foothills twenty-five miles away. Air traffic is light at 9 P.M. and he has ample time to weave a DF steer into his slowing volume of weather reports, clearances and airport advisories. The man switches on a scope to capture a slit of light that signifies our airplane. In the next ten minutes he will attempt to lead the glowing trace to Dutchess County Airport at the center of his screen.

"Transmit for five seconds," he orders. As I hold down the mike button, the controller spots our blip and calls for a compass heading of 330 degrees. A glance ahead through the murky windshield proves how reassuring it is to know that a DF steer is just a radio call away. Moments after take-off we had surrendered any sense of time or position and allowed the airplane to wander in the sullen twilight. The voice in the radio every few minutes would certainly comfort a bewildered pilot tossed about by disorientation.

"Transmit for five seconds and turn to a heading of 320 degrees." Our trace was being prodded off course by the wind. To correct for drift, the controller issues three more course changes in the ensuing minutes to keep us tracking to his station. Finally, the voice breaks its official tone to utter the last instruction:

"Look out the window. You can spit at the airport!"

Neither of us aboard can see the field. The DF steer has been executed with such delicate accuracy that our plane arrived directly overhead and blotted out the airport. I drop a wing to catch sight of the mile-long lights that edge Runway 24.

The landing approach to this airport, surrounded by hills and high towers, must be handled with caution. Before the flight we learned that airport elevation is 165 feet, but a scant five minutes' flying time from the field the obstructions loom as high as 1,740 feet (MSL). To reduce any risk, the approach will be along the sloping VASI lights that lead to the threshold of the active runway. After a letdown to 3,000 feet and a distance of eight miles, we see the VASI lights straddling the runway and reduce power for the descent.

If VASI lights are difficult to find while approaching an airport, it's helpful to start the search at the rotating beacon. When it's seen, align the airplane with the runway heading, even if the field or runway lights are not yet in sight. This maneuver places the runway in a known direction (on, or parallel to, your line of flight) and helps separate it from other lights below. As you near the airport, scan the beginning of the runway area (look for the green threshold lights) and the VASIs should appear. They are probably white this time because an initial approach is usually from an altitude above the visual path formed by the lights. Our descent into Dutchess County Airport ends uneventfully after the plane glides down the white-red beams to touchdown.

POU–PWL

MC—70°
Dist—19 sm
ETE—9″
Obstructions—1,600′

The northwest wind blows briskly on take-off from Poughkeepsie at 9:26 P.M. This leg of the journey will carry us over territory that bears few yellow areas on the sectional chart, a warning that almost no masses of ground lights can serve as checkpoints. If the horizon fades in poor visibility, it will be difficult to fly this leg solely by looking out the window. The plan is to fly nineteen miles to Pawling VOR, then navigate by dead reckoning to a tiny airport called "Johnnycake." We will circle its rotating beacon to establish a definite checkpoint before the approach into Bradley International. The city of Poughkeepsie, last bright patch below, disappears behind the left wing as the plane nears PWL to begin the dark leg of the journey. As if commanded by the terminal forecast, a deck of 5,000-foot clouds floats across the route to extinguish the moon's pale illumination. Pawling VOR flicks the TO-FROM flag in the receiver, a thumb punches the stopwatch, and the airplane penetrates the wall of black to the east.

PWL–Johnnycake

MC—102°
Dist—30 sm
ETE—14″
Obstructions—1,800′

The few lights ahead on the rural countryside slue from right to left across the windshield to tell the persistence and strength of the northwest wind. If our track is left uncorrected, drift will carry the plane many miles south of the next checkpoint. A swing of the nose toward the wind's source attempts to freeze the sliding lights and test a new compass heading. If the track is true, the checkpoint will appear when the stopwatch ticks off fourteen minutes.

There is a haunting quality to flying over dark terrain in a single-engine plane. As signs of the earth withdraw, the eye wanders to an unseen horizon to measure the miles to a distant glimmer that communicates only that it is there. Conversation between pilots deadens in the narcotic effect of unremitting blackness. An unheard signal summons the gremlins to play—when a change in wind direction alters the timbre of the propeller, the sound clatters in awesome intensity. The ammeter needle flirts with the "Discharge" arc to taunt that it can empty the battery. The throttle cable feels loose, there's slop in the controls . . .

Somewhere ahead a dash of light inflates to a triangle and brightens. A town is coming to life on the dark sphere. Its outlines, when compared with the yellow contours on the sectional chart, identify it as Torrington. But it's too far to the right. A row of miniature headlights moving on a north-south line confirms a lone interstate highway that grazes Torrington. With the checkpoint identified, we fly a careful compass course toward tiny Johnnycake and strain to see a rotating beacon. Nothing appears. The stopwatch has two minutes to go. Did a change in wind velocity slow the Cherokee's ground speed? Maybe poor drift correction led us astray.

"I see a strip below," my friend says, looking over his right shoulder. I turn the plane in a triumphant circle over the field.

"But is it the right airport?" he challenges. With few landmarks at night to verify a small field, it's easy to make a mistake, possibly with dire consequences, and attempt a hazardous landing at the wrong destination. Confirmation of the field will come several minutes later.

Johnnycake–BDL

MC—70°
Distance—20 sm
ETE—10″
Obstructions—1,800′

Bradley International Airport sprawls twenty miles to the northeast. Its high-wattage beacon would easily be seen at this distance on another night, but the nebulous air holds visibility to about ten miles. Only an indistinct glow in the east hints at the complex of towns that cluster around the state capitol at Hartford. To avoid tangling with heavy jet traffic into Bradley International we will listen on ATIS (Automatic Terminal Information Service) to picture the traffic patterns. At the last checkpoint, Johnnycake Airport, a call to Bradley's radar facility (listed in AIM) might bring approval for a straight-in approach to Runway 6 from about fifteen miles out. In this approach, we could make a precise track on the localizer beam, and also fly under radar surveillance. The two tactics will cause least hazard to ourselves and other aircraft operating into the field.

"Radar contact 15 southwest. Maintain heading of 90 degrees and intercept the localizer. If unable to contact the tower at the outer marker, break off the approach to the right."

The voice of Bradley Approach Control responds to our request for a vector from our present position, now pegged as Johnnycake, to the localizer for a straight-in to Runway 6. Even as the radar operator speaks, the lights of another airplane blink at our altitude about twenty miles away. Its frozen position in the windshield means the plane is converging with our flight path. Clearly, this airplane is also on an intercept to Bradley's Runway 6. But our call to approach control removes any apprehension about a midair collision. The other plane, a DC-8, is permitted to proceed on course; the controller knows the jet's speed will easily outpace a Cherokee in the approach to the localizer. (The airliner, though, would reappear as a factor on the ground.) The remaining miles to the field can now be completed in comfort. According to regulations, our approach could have ignored the radar facility and radio contact made directly with the tower five miles out. It was much safer to have a watchful eye on the radar screen to head off any aerial conflict. At eight miles from touchdown, green-white fingers of Bradley's beacon sweep our nose and draw the Cherokee toward the airport area. Then a tiny, glowing cruciform expands into the approach lights, its long shaft pointed at us, the crossbar angling against the runway's yet invisible threshold. It is still difficult to visually estimate distance to the field because the approach lights emit a conflicting perspective of high brightness and small size. We will hold 2,500 feet until a definite checkpoint guarantees a clear descent.

A swing of the ADF indicates that point. Tuned to Bradley's compass locator, the dipping needle proves that five miles remain to touchdown.

"Caution, wake turbulence, landing DC-8," warns the tower. (It's that jet seen aloft earlier.) The airliner's wake will create no hazard to us, because enough time will elapse before touchdown for the vicious wingtip vortices to dissipate.

Time passes agonizingly slow on a straight-in nighttime approach to a well-lighted runway. With little movement outside, the abyss between airplane and field shortens by imperceptible degrees. But approach lights ultimately strengthen and cleave into touchdown lights. Rapidly, the airport illumination flares, rises to meet the approaching aircraft, then streaks by in dazzling testimony to the airplane's true speed and motion. A swoop over the threshold and we are on the runway at Bradley, following the ground controller's voice through the sea of psychedelic lights that washes over a big airport.

BDL–MMK

MC—210°
Distance—21 sm
ETE—10″
Obstructions—1,288′

After checking weather, our take-off from Bradley will begin an ADF-tracking leg to a radio beacon (MMK) at a small airport in Meriden, Connecticut. The leg will test the electronic vagaries of a nighttime ionosphere and any blurring effect on the ADF needle. Invigorated by the chilly air, the Cherokee lifts high over the luminous city of Hartford in a long sweeping turn to the south. A wind on the tail promises a welcome push for the overwater crossing still thirty minutes away. A twist of the ADF dial draws a cacophony of crackles and hisses from the radio speaker, then the faint beeps of MMK in Morse code. But the ADF needle flops over the dial in a senseless search for an ADF signal still too weak for tracking.

It is 10:45 P.M. The stopover at Bradley International had exposed the first embarrassing glitch of our flight. Back on the ground, I had complained to the specialist in the FSS office that the rotating beacon at Johnnycake Airport was out. It was hardly a serious threat to the safety of air navigation, but the man dutifully spread his chart and looked for the star symbol that marked a beacon. It had none. To prove I wasn't wrong, I checked the date on his chart and, with a flourish, pointed out that his copy had expired several months ago. By now, three FAA men huddled over the missing beacon. One specialist withdrew to another room and telephoned the airport manager at Johnnycake Airport. A testy lady on the line replied: "Are those fellers flying with a 1901 chart? That beacon has been out for twenty years!"

I stalked out of the Flight Service Station and grumbled about the inability of the U. S. Government Printing Office to keep its symbols straight. I surely saw that star while planning the trip on the ground.

The final answer appears during climb-out from Bradley on the leg toward Meriden. I poke the flashlight beam on a sectional chart—the latest edition—and probe for the twinkling star of an airport beacon imprinted over Johnnycake. There is no trace of the symbol. I had looked at the wrong airport while flight-planning that afternoon. It was easy to rationalize that any flight—from jumbos down to J3s—suffers at least one mandatory goof. But what if our trip was aimed at the desolate blackness of the Maine woods? A misplaced beacon could multiply into a minor crisis.

The ADF needle halts its aimless circle around the azimuth card and stabs a steady heading. MMK is radiating a vigorous signal that slices through the gibberish of "night effect" to draw us on a sure course toward the coast. I ponder an engine failure along this leg—not so much the prospect of a blind landing in the dark but a year-old newspaper headline about a downed aircraft in this part of the state. Rescuers didn't find the wreckage for three *days*. The account ended with the usual remark about plane crashes and calamities at sea: "Rescuers stopped searching because of darkness." Perhaps the urgency about filing a flight plan at night holds a hollow promise. Would the Civil Air Patrol trace over the hills and light up the crevices with searchlights to find us? Probably not. We invoke another form of nighttime insurance—one radio continuously tuned to a ground station known to be within hearing range. It might be a tower, the emergency calling channel 121.5 MHz, or a favorite frequency, the closest Flight Service Station. Tuned to an FSS, we would also hear the routine chatter of weather reports and voices of other pilots headed toward exotic destinations a thousand miles away. For the rest of this trip, we would monitor New York Radio, located ahead on Long Island. If the engine stopped, I could reach the mike, shout a Mayday and position report while barely out of cruise altitude.

It is almost time to call New York Radio, anyway. At 10:57 P.M. the ADF needle smartly reverses to announce an arrival over Meriden beacon and the imminent water crossing. The Cherokee flight manual said the plane would glide less than eight miles from an altitude of 4,500 feet, and calculations showed this altitude would expose us to a chilling swim of several miles to shore if the engine stopped somewhere near the middle of Long Island Sound. The risk had been taken on dozens of earlier flights across that body of water but always when the water temperature was survivable. Those hops were also in daylight, when scores of boats would catch sight of a crippled airplane and rush to the rescue. But a night crossing was foolhardy unless altitude allowed enough margin to glide to shore from any position. If the forecast of 5,500-foot clouds prevails we will alter course and traverse a narrower neck of the Sound.

New Haven on the shoreline grows luminous in the windshield. Beyond, the galaxy of city lights abruptly drowns in the black water running to infinity. High above, a scattered layer yellows the half-moon and softens its edges. An indistinctness in the air obliterates objects more than ten miles in the distance. But somehow the 5,500-foot clouds have effused into wisps of gray just under the stratosphere at 25,000 feet. With no altitude restrictions, the Cherokee climbs strongly under full throttle. We will make the crossing at a secure height of 6,500 feet.

MMK–RVH

MC—208°
Distance—42 sm
ETE—20″
Obstructions—1,549′

"New York Radio, this is Cherokee 6654J." The receding shoreline warns us to call the Flight Service Station and activate an overwater watch. We tell the man our track will follow a radial direct to Riverhead VOR and await his reply.

"Call me when you're halfway across," he asks. A click of my stopwatch commences a five-minute count to mark the midpoint of the crossing. Pitch-blackness creeps over the left window like a velvet curtain. Off to the right, a feeble moon tests a reflection against the water's surface a mile below. A glance at the artificial horizon warns that the right wing is slightly low, a subtle change in attitude hidden by miles of mist between us and the natural horizon. I try to hold the VOR course exactly, imprisoning the needle within its circular cage. Now ten miles from shore, cut off from the lights of either coast, there is nothing to grasp but the knobs and dials within the tiny cosmos of the cockpit. Motion beyond the windows ceases.

Could some uncanny cabal of wind and error carry us to the open Atlantic? It is a foolish fear. Maybe not—since the Sound widens into the Atlantic Ocean twenty minutes to the east, only ten minutes more flying time than our overwater leg. If that forecast cloud layer of 5,500 feet sneaks below us, it might conceal the shore and obscure a landfall. My speculation is interrupted by the memory of a bizarre incident that happened three years ago in this area. It was also late evening and in the same type airplane! A young man with a couple of hours of flight instruction and a history of mental illness had stolen a Cherokee and flew it in the dark until it ran out of gas. The plane touched down 140 miles out in the Atlantic and flipped on its back when the nose wheel dug into the water. With the incredible chance that often rides with tragicomic events, the man was plucked from certain death by a Coast Guard helicopter that had followed him out to sea.

Something blossoms silently in our right windshield. A glimmer at first, followed by an exploding circle of light on the water. It collapses in the dark, then blooms brightly again.

"New York Radio, Cherokee 6654J is halfway across." The lone light below is a flashing marine buoy we had spotted earlier on the sectional chart. As a pilotage checkpoint, it proves we are on course. The buoy hints also at rising visibility because its beams are clearly seen more than a dozen miles away. A thin rim of light ahead thickens into the coast of Long Island.

RVH–ISP

MC—248°
Distance—11 sm
ETE—5″
Obstructions—646′

Final touchdown is somewhere over the next eleven miles in the dazzling maze which surrounds Islip Airport. The field is a favorite stop for instructors taking students on their first night checkout. Finding the field amid a skein of highways and suburban developments is a test of careful VOR tracking and timing. There are no approach lights to send up a guiding cross of illumination to the runway. I hit the stopwatch at Riverhead VOR and turn to the 248° heading demanded by the flight log. Trap the VOR needle on its dial and Islip Airport should emerge from the miasma in less than five minutes.

It is almost 11:50 P.M. The last traces of haze unnoticeably evaporate into the night atmosphere. A quarter-million miles away, the moon suddenly enjoys a crisp semicircular outline in the clearing air. Under our wings the amazing vitality of big-city suburbs at midnight seethes on the highways. Squeezed between the black waters of the Sound and the Atlantic, Long Island sends a blazing belt of light to the western horizon.

"There's the field," my friend says quietly. Green and white shafts tug at his eye.

I had been marveling at the unusual brilliance of the evening. Fifty miles ahead, mammoth towers of the World Trade Center on Manhattan scrape against a pure black sky.

"54J is cleared to land straight in on Runway 24," drones Islip Tower.

The Cherokee's wheels kiss the dark asphalt and spin toward a yellow stripe that gracefully curves off the runway. A parade of blue taxiway lights whips past our windows to salute the end of a perfect night flight.

Appendix

INJURIES

	FATAL	SERIOUS	MINOR	NONE	TOTAL
Pilot	208	80	97	339	724
Copilot	14	1	2	6	23
Dual Student	—	2	7	15	24
Check Pilot	—	—	—	—	—
Flight Engineer	—	—	—	—	—
Navigator	—	—	—	—	—
Cabin Attendant	—	—	—	—	—
Extra Crew	1	1		1	3
Passengers	240	93	97	437	867
TOTAL	463	177	203	798	1641
*Other Aircraft	—	—	—	59	59
Other Ground	—	2	2	2	6
GRAND TOTAL	463	179	205	859	1706

Involves 722 Total Accidents
Involves 213 Fatal Accidents

*Injuries carried opposite other aircraft are injuries occurring in aircraft that are not part of this subject tabulation, but were part of the total injuries involved in collisions between aircraft.

105. Injuries, accidents which occurred at night involving single-engine aircraft. U. S. General Aviation 1969–70.

Involves 722 Total Accidents
Involves 213 Fatal Accidents

	FATAL ACCIDENTS			NONFATAL ACCIDENTS			ALL ACCIDENTS		
BROAD CAUSE/FACTOR	CAUSE	FACTOR	TOTAL*	CAUSE	FACTOR	TOTAL*	CAUSE	FACTOR	TOTAL*
Pilot	194	40	195	430	30	431	624	70	626
Personnel	2	3	5	47	9	55	49	12	60
Airframe	2	—	2	3	—	3	5	—	5
Landing Gear	—	—	—	10	5	14	10	5	14
Powerplant	8	—	8	48	—	43	56	—	56
Systems	—	—	—	7	4	11	7	4	11
Instruments/Equipment and Accessories	1	2	3	1	3	4	2	5	7
Rotorcraft	—	—	—	—	1	1	—	1	1
Airports/Airways Facilities	1	2	3	10	46	56	11	48	59
Weather	25	103	126	31	55	84	56	158	210
Terrain	8	11	19	54	35	89	62	46	108
Miscellaneous	2	—	2	21	3	23	23	3	25
Undetermined	14	—	14	10	—	10	24	—	24

The figures opposite each causal category represent the number of accidents in which that particular causal category was assigned.

*If an accident includes both a cause and related factor in the same causal category, the accident is represented once under the total for that category.

106. Cause/factor table. Single-engine aircraft accidents which occurred at night. U. S. General Aviation 1969–70.